Another William Kennedy has made their mark.
Kept my interest from start to finish. Great read. Looking forward to
the next book. Greg Perkins, author of the The Announcers. March
2018.

**An exciting novel making this reader hope Kennedy's second
novel is published soon.**
Jonathan West, MD – First Kill by William Kennedy is an exciting
novel of the recruitment of a young agent by the CIA. The agency
appeals to his patriotism to participate, as DB Cooper, in hijacking
an aircraft to convince the American public to support an increase
in airline security, an interesting interpretation of American history.
The seduction continues by the CIA financing West's medical
education at Yale and assigning him light intelligence duties to
collect information on Iranian students in pre-revolutionary Iran.
His first kill is carefully crafted as West fears the Iranian student
will plunge Iran into a serious revolution, impacting world oil
markets, and might harm his fiancé. Frank E. Hopkins, author of
Abandoned Homes: Vietnam Revenge Murders, The Billion Dollar
Embezzlement Murders. September 2016.

Real story behind DB Cooper
Great first book by Dr. Kennedy, love how he used DB Cooper in this
book. Clever! I've heard that his next one is coming out soon and I
look forward to it. Jimmy D. March 2018.

Kennedy has found the right balance...
Kennedy has found the right balance between a fast paced creative
and stimulating plot and sufficiently realistic character development
that creates the tension that readers of the genre look for in novels.
It is surprising that he has found this balance in the first work and
I look forward to reading other that he creates. Bert Spilker, MD,
author. March 2018.

Page turner

Kennedy present the main character, Jonathan West, with something so inconceivable that gradually springboards him into a world far different than his chosen profession. There are many interesting directions to take this story and the author does not disappoint. It is well written to entertain and provoke thought. Well done. Hoping there is a sequel! Victoria. August 2016.

Praise for Morally Gray

He' back!

Our favorite murderous MD, Jonathan West, is up to new adventures with the CIA. When the Shah of Iran neared the end of his reign, the focus was on his successor but the problem, as seen by the CIA, was the still living Shah. Author William J. Kennedy provides a credible tale of solving the problem via Jonathan West MD's medical skills and international connections.

Follow along the serpentine trail between New York City and Tehran as medical mayhem is melded with patriotic righteousness to solve the problem of the Shah. Walter F. Curran, author of the Young Mariners series. June 2019.

West is Back! Enjoy the Trip

Jonathan West is back! West, the doctor who rejects "do no harm," is the man behind historical events making them not what they seem. This time, the young doctor is married, starting a family, but the CIA keeps him from settling down.

In Morally Gray, his job is to make friends with Iranian operatives in New York. This is the waning days of the shah's rule in Iran, a man placed in power by the CIA and a friend in the Middle East for the US. As West gains the confidence of the right people, he works his way through the power hierarchy to the side of the Shah.

I won't spoil the outcome, but what happens to the Shah involved West. Enough said.

Kennedy writes inside the head of West and lets you see his world from his point of view. It's a well-paced spy novel. You get to know the man as a friend and wonder sometimes if that should bother you.

Get the book, settle back and enjoy the adventure. Jackson Coppley, author of Leaving Lisa and The Code Hunters. May 2018.

Another good story!
Another interesting story and a lot more detail on the main character. Enough plots and intrigue to keep one wondering what was next. The conflict between conscience and duty was riveting and insightful. I would recommend this to anyone interested in adventure, travel and James Bond tribulations and intrigue. Pete Gendron, May 2018.

THE PENTAGON YEARS

WILLIAM J. KENNEDY

Other Fiction by William J. Kennedy
  First Kill
    Previously published as Jonathan West, MD – First Kill
  Morally Gray

This book is a work of fiction. Names, characters, places, and incidents are products of the author's imagination or are used fictitiously. Any resemblance to actual events, locales, or persons, living or dead, is coincidental.

© 2019 William J. Kennedy
All rights reserved.

ISBN: 978-1-7334949-0-8
Library of Congress Control Number: 1733494901
Blackpool Creek Publishing
Rehoboth Beach, DE

This book is dedicated to the men and women, civilian and military who serve our country.

Acknowledgments

I would like to thank once again the members of the Rehoboth Beach Writers Guild, a group of very talented writers who are a constant source of encouragement. Special thanks to the Gray Guys Group for our regular critique meetings that challenge and motivate me. The Wednesday night coffee and gab group of Cindy Hall, Ilona Holland, Frances Oakes, Mary Ann Hoyt, Walt Curran and Crystal Heidel was most helpful in helping me put the finishing touches on the book. My beta readers, Frank Hopkins and Tom Hoyer, provided the luster on an almost polished manuscript. A profound thank you to my editor, Robert Bidinotto. And to Crystal Heidel of Byzsantium Sky Press, thanks for another great cover. Thanks to Nick and his team at WORD-2-Kindle for the final formatting.

CHAPTER 1

"...a date with Sara Dipity..."

Monday, July 21, 1969

I ducked behind the supply room to sneak a smoke after morning chow. Didn't really have time before formation, but I figured, What the hell! What's the worst thing they could do, send me to Vietnam? I'm going to go there anyway. Vietnam couldn't be much worse than where I was now: Fort Dix, New Jersey, in the summer, probably spending this last summer of my life in the Army Clerk Typist school learning to file. Yes, file "A" before "B" except after FTA, as in "fuck the Army."

I finished my L&M, field-stripped it, rolling the tobacco from the finished smoke into the breeze and put the filter in my pants pocket. As I came around to the front of the barracks complex and joined my company, one of the Advanced Individual Training (AIT) sergeants walked toward me, and I prepared myself for an ass-chewing for being late.

"West, where the fuck you been?" he asked. The tone was friendly despite the words he used. "Fuck" had become the noun, verb, adverb, and adjective of choice in all conversations. If you didn't use it, you were speaking a foreign language.

1

"Stomach was gurgling," I lied. "Thought I was getting the trots." The sergeant didn't respond, just looked right at me, through me, reading my eyes and my thoughts. "I was grabbing a smoke," I confessed.

"You missed the announcements. They called out ten guys' names to go over to Post Personnel. One name was yours. Sounded important."

"Maybe the war is over, Sarge, and they're sending me home. Kind of like the unions—last in, first out."

"Ain't likely, so get over to CQ and join the rest." CQ, I remembered, Charge of Quarters.

I double-timed the two hundred yards to the company office area and joined nine other guys waiting around. A clerk checked names against a list he had on the standard Army-issue clipboard. I walked over and said, "Do I have to check in aga in?"

"Where you been?" he asked.

"First Sergeant called my name at formation and told me to get over here."

"You're on the list, but you're not marked present."

"What do you believe, the list or your eyes?" I said. The corporal looked at his list and then at me. I felt I could take some liberties with him. He had been my supervisor last week as I painted an empty classroom. He was a decent sort, taking coffee and smoke breaks with me. One lunch, he felt like pizza. He borrowed a car and got one at the Post Exchange, brought it back, and invited me to share it with him. "List says I'm not here, but your eyes say I am, bigger than shit." He corrected the "error" by making two quick check marks next to my name on his list.

"Wait here," he ordered, satisfied he had done his job of checking off the correct number of boxes.

So, I waited along with the others. Waiting for what, I didn't know, but I had learned to wait during my first ten weeks in the Army. Hurry up and wait, isn't that what they say the Army is all about? It's true. I ran over here to wait for something.

An olive-drab school bus pulled up, and we got on. As a trainee, we rode nowhere we could walk. This ride was less than a mile, a mere fifteen-minute walk. What's up? Why the special treatment?

The bus stopped in front of a sprawling, one-story, sand-colored brick building. It looked like the new schools that had sprung up in suburbs all over the country. Red brick was a thing of the past. Sand-colored brick was in. Conclusion—this was one of the newer buildings at Fort Dix.

Inside the double glass doors, any similarity to a suburban school disappeared. A waiting area thirty feet wide by twenty feet deep. The wall opposite the door was a solid panel extending halfway up from the floor, with the top half a glass partition to the ceiling. At one end of the half-glassed wall was a door, while the other end had a window-like opening. A soldier stood behind the window.

"Are you the guys from the clerk school?" he asked.

One guy stepped forward and responded, "Yes, Corporal. First Sergeant told us at formation this morning to report to Post Personnel."

"When I read out your name, just say yes or here or present," he said. More confusion. They gave us a choice. The Army hadn't given us a choice about anything. Was this a trick? I stuck with the standard "present," to which I added "Corporal," just to be safe. Then, he told us to take a seat. Four of the guys parked themselves in the four chairs available, and the rest of us stood.

Standing in the waiting area gave me a chance to look around, beyond the glass partition, into the working area of this building, which was the size of a football field. There were offices around the outside wall, but the interior contained wall-to-wall desks.

The door to the back opened, and a Major stepped into the waiting area with a stack of folders, presumably our 201 files.

"Gentlemen," he began. The first time someone had addressed us this way! He continued, looking from soldier to soldier as he spoke. "The war in Vietnam, especially the Tet Offensive, has put unanticipated demands on the U.S. Army."

Shit, I thought. They're running short on soldiers for Vietnam, and they're going to send us over there today. Dozens of thoughts ran through my mind, all competing for space, time, and the top of the list. The one that won was, Can I escape to Canada before they ship me to Vietnam?

Before I had to decide on a plan, he continued. "Because of the increased activity in Vietnam, our systems for managing personnel have been overloaded. Just look behind the glass here in this building. Many of the soldiers you see were in Clerk School last week, and now they're sitting at desks getting 'on the job' training. The same situation exists in the Pentagon. Civilians do most of the work at the Pentagon, but the work has increased so dramatically they can't handle it. They've asked us to screen all the soldiers in our clerk school to find candidates to go to the Pentagon and help. You've been selected because you are in Clerk School, because you have college degrees, with some having advanced degrees, and because you can type. We've done a preliminary security and background check on you, and today I'll interview you. If selected, I'll notify you before you leave the building. You'll then be given two days to clear post and leave for the Pentagon on Thursday. Are there any questions?"

There were no questions! Our gaping mouths were busy catching flies, and by the time our mouths closed, he called the first name.

The first guy came out looking dejected. He'd been in there for less than ten minutes. Before we could talk about it, the Major stood at the door in the half-glassed wall calling for another interviewee, Dexter W. Green.

My mind drifted. If I got this assignment, I'd be in Washington rather than ten thousand miles away on the other side of the planet.

Dexter came out with a smile on his face and gave us a thumbs up as the major called out "West."

Raising an arm to identify myself, I moved forward and followed him.

The Major's office was small and cluttered and seemed darker than the interior offices, despite the double exterior windows.

The Major began. "West, you seem to fit the bill. You whizzed through Clerk School and you can type." Getting through the eight-week Clerk School in four days may seem like whizzing to the Major, but to the rest of the educated world, not so much. When the draft boards scraped the bottom and sides of the barrel, they picked up a lot of college graduates. Many made it to the clerk school, where the emphasis was on learning to type; but while learning to type, the students also learned to file, do basic arithmetic for the morning report, and do the mail. If you could type, the alphabet and arithmetic were a snap, so the college guys sat around. The Army recognized this and made it self-paced. I was one of the slower ones, with a four-day completion. One guy finished before noon on the first day. He spent the rest of the first week painting the classroom next door.

"What do you think about the Apollo 11 landing on the moon yesterday?" he asked.

"No shit!" I responded, with true surprise. "Sorry, sir. I hadn't heard that. Wow! A man on the moon. Sorry, sir, but I haven't had much access to the news since I got to Fort Dix. Guess you could say I'm out of touch. I'll have to make the effort to get current again."

Not looking up from the file on his desk, he stated, "With your Harvard degree, I'm surprised you're not an officer." Then he added, "Or headed to Fort MacArthur as an athlete in Special Services."

"Sir, I never considered being an officer, once I learned it meant serving for three years and being subject to being called up at any time." He looked up from my file. I had struck a sore point. "As far as playing sports for the Army, I enlisted for three years to play hockey, only to learn once I got here the Army doesn't have a hockey team."

The Major signed the document in front of him with a flourish, extended his hand, and told me my company clerk would have orders for me by noon to start clearing post.

"Nothing fancy for the trip. A Trailways bus will pick you up at the Post Reception Center Thursday morning. Congratulations, Private West."

"Thanks, sir." And that was it.

I left through the door in the half-glass wall with the Major behind me. As I entered the reception area, he called out the next person, "Kennedy."

None of the other guys who had been interviewed before me were in the reception area, so I went over to the soldier at the window.

"I just got interviewed for the job at the Pentagon. What's next?"

He looked up and asked, "Did you get the job?"

"Yeah. I leave Thursday. Looks like 'Peace with honor' needs some help pushing paper."

"Congratulations, West. You can walk back to your company area, or you can wait outside for the bus that'll be back in about a half-hour."

"I'll wait."

I walked outside awaiting orders assigning me to the Pentagon, to leave by public transportation on Thursday, the day after the day after tomorrow—or as I would count time from now on, two days and a wake-up. Did I feel good? You bet I felt good! I was special! I was going to the Pentagon. The wonderfulness of myself was only exceeded by my profound sense of humility! Thank you, Dad, for making me take typing in high school.

Outside it was still Fort Dix, New Jersey, still July, and still hot, but it seemed nicer. The sky a little bluer, a little breeze, and the smell of something pleasant in the air. Not flowers, or a steak grilling over an open fire, just something nice about the air.

I lit up an L&M, my second smoke of the day. Even the smoke tasted better. I waited for the bus because, when given a choice, I'd rather ride than walk. As I field-stripped my cigarette, the guy who interviewed after me, Kennedy, came out. Smiling, he walked over.

"I'm going to the Pentagon," he announced.

"Me, too. Jonathan West," I said offering my hand.

"Jim Kennedy," he replied.

"Not much of an interview," I said.

"Not much. Just two questions. Did I hear about the moon landing, and did I hear about the accident on the Cape?"

"Yeah. He asked me the question about the moon, but not the Cape."

"I guess they found out about me doing the background check. It impressed the hell out of me. I called home on Sunday night, my wife and my parents. My wife told me my parents weren't home, and I should call my aunt Bobbie. She told me my sister had been in a motorcycle accident on the Cape, and Mom and Dad drove down to be with her."

"He asked you about that?"

"Yeah, he wanted to know how I knew about it when I hadn't heard about the moon landing. When I told him I found out when I called Bobbie's house, he got excited. Said I was on my way to the Pentagon, and if there was anything he could do for me, let him know. Even gave me his card. Look."

I looked at the card, a civilian business card. The major was a vice president of a bank in Chicago. A lot of Army Reserve officers got called up for an extra six to twelve months active service. He must be one.

"So, Kennedy, where are you from?"

"Worcester, Massachusetts. He made something of that, too. Something like Kennedy from Massachusetts."

"You related to any of the politicians?"

"No, I'm not even a Democrat."

We enjoyed another smoke and more idle conversation until the bus came.

The bus dropped me off in the company area a little after ten o'clock. Kennedy was in a different company and stayed on the bus. Once back in the area, I tried to lie low until noon chow. If any of the permanent cadre saw me doing nothing, they'd find some worthless thing for me to do. I stayed out of sight for the rest of the morning and headed over to the mess hall for lunch.

I was one of the first through the serving line. After I picked a table and settled down, the company clerk saw me and hustled over before getting in line.

"West. Where the fuck you been? Everyone's looking for you, even the Captain."

Looking casual as I ate my grilled cheese sandwich, I said, "I had a meeting in Post Personnel all morning. Just got back."

"You got orders for DA. You're shipping out on Thursday."

Surprised, I dropped my sandwich. "No, you got it wrong. I'm going to the Pentagon."

"You moron. DA is the Pentagon—Department of the Army, DA."

"Right," I said and continued eating. During my short time in the mess hall, two Sergeants from the company cadre stopped by to congratulate me on my assignment. I learned I was the only one from my company going to the Pentagon.

The rest of the day, and the next two days, were anything but the routine I had gotten used to. I didn't have any work assignments: no guard duty, no KP, and no painting. The Company Commander asked me to stop by his office for coffee and a chat in the afternoon. Tuesday, they sent me to the post supply to get new khaki uniforms, new Class As, and new shoes. Once I received the new uniforms, I strolled to the tailor shop for a fitting. They didn't want to send their Fort Dix soldier to the Pentagon looking like a "sad sack." After a casual breakfast on Wednesday, I went to the barber shop before picking up my altered uniforms.

It was the first time in twelve weeks the barber asked me how I wanted my hair cut.

"Nothing off the top. Just clean up the back."

CHAPTER 2

"One Mystery Solved"

Thursday started at 0430, still dark, but the thought of ending the day at the Pentagon brightened my mood. The Supply Sergeant had been on CQ duty all night and gave me a ride to the bus station, continuing the sense of good will from the cadre. When I arrived, I saw only two other soldiers waiting beside the idling Trailways bus, Green and Kennedy. Why are there only three of us on the bus today when they had interviewed ten?

I said hello to Kennedy and said "Jonathan West" to the other.

Nodding, he replied, "Dexter W. Green," and extended his hand. I dropped my duffle bag next to theirs and shook it.

"Ready for the adventure?" I asked, not expecting a response.

"Into the unknown," replied Kennedy.

Dexter W. corrected him. "Actually, we're off to Fort Myer, Virginia, and our duty station will be the Pentagon."

"Fort Myer? The one at Arlington Cemetery?"

"Yes."

And that was the end of our discussion until the bus driver opened the baggage compartment and told us we could toss our bags in.

"You have to change buses in Philadelphia, and I don't take any responsibility for your hangers."

Yes, the hangers. Wooden hangers hooked on the outside of our duffel bags. The Army issued us three, but we needed a dozen to hang uniforms, both Class "A" and fatigues. Meaning, we had to buy the rest. Eleven were strung through the webbing of the duffel bag that served as a shoulder strap. Why eleven? When clearing Fort Dix, the Army made each of us turn in his field jacket on a hanger even though it didn't come on a hanger. Somebody was collecting the hangers and reselling them to the next crew that came through.

None of us knew each other before we met up at Post Personnel on Monday. Each of us had done his Basic Training at the same time, but none in the same company. Because of the new curriculum of self-paced learning in Clerk School, we had never gone to a class together. While all of us were "college boys," a designation hung on us by those high school drop-outs who served as our cadre during Basic Training, we shared few other things. Before the Army, we were fresh out of college and starting careers or graduate school. Common dreams included getting an education, getting a job, getting drunk, getting laid, or getting married. Similarly, we shared the dream of not being drafted into the Army.

Our lives and dreams before the Army were replaced with dreams of avoiding Vietnam, staying alive, and someday returning to the lives we enjoyed before.

On the bus, leaving Fort Dix, I smiled. A bus had brought me into the Army. At the time, it was a low point, and I associated it with riding the bus. As time in Basic moved on, I remembered the bus ride fondly and set riding the bus again as a goal. The first thing I remembered about those early weeks was walking—the Army called it marching. The worst part was walking in the sand at the side of the road, a perfectly good road with no traffic. A wasted road because we walked in the sand on the side of the road. We aspired to walk on the road. Sometimes, a truck drove by with soldiers riding in the back, something else we desired. The thought of a ride in a private vehicle, or driving our own car, was out of the question. The ultimate aspiration was to be back on an air-conditioned bus, preferably leaving Fort Dix.

I thought of the aspirations we shared now, dreams almost as identical as the uniforms we wore: uniforms of heavy tan khaki material, which made them durable and sucked up starch when subjected to an Army laundry. They looked good fresh off the hanger, but they weren't made for hot, humid days spent sitting on a bus. We got these new dreams from the Army, and like the uniforms, the dreams were wrinkled. We might have avoided Vietnam, but we were still a long way from returning to the lives we had before.

In Philadelphia, we exchanged our travel vouchers for real tickets. Enjoying the return to this small slice of civilian life, I got a cold hot dog, a warm cold drink, and a newspaper, before I stood in line with the civilian passengers for the final leg of our trip to Washington. Once again, we placed our duffle bags in the gaping hole that served as the luggage space under the silver big brother of the yellow buses we took to school as kids. The bus wasn't crowded, so we had a choice of seats, but we gravitated to the back.

Picking the left side of the aisle, I took a window seat. With the hot July afternoon sun on our right as the bus drove south, I preferred any coolness the shaded side of the bus would offer. The seats were a coarse blue fabric but felt better than anything the Army had provided during our short association. The air conditioner worked a lot better than the one in the bus station. Dexter W. and Kennedy sat behind me, each of us having two seats to ourselves. Comfortable, I settled in for the trip and opened the newspaper.

The simple pleasures of reading a paper returned. There was a tactile and olfactory reaction as soon as I unfolded it. Auditory stimulation kicked in when I snapped it open to examine the full front page. The mere act of reading this paper had stimulated four of my five senses. Dare I lick the page to engage the fifth and final?

Stories about the moon landing dominated the front page. Our national pride compelled us to boast about fulfilling JFK's promise of having a man on the moon before the end of the 1960s. I read every word, moving to the continuation inside where I saw a headline on page three. "Kennedy Accident on Cape Cod Still a Mystery." My head swiveled to the right and rear, seeking Kennedy. Was his sister a celebrity to warrant a story a week after the fact? He had taken a window seat two rows back and I couldn't see him, so my eyes returned to the story. By the time I finished the second paragraph, it became clear the story wasn't about his sister.

Sliding out of the seat with the paper in hand, I moved back to his seat. His head rested against the window and his eyes were closed.

"Are you awake? I asked.

"Just dozing. Forgot how nice it was to do nothing in the middle of the day."

"Have you read the paper?" I said, noticing the paper on his lap.

"Just the front page. Lots of stuff about the moon landing."

14

"Tell me again about your conversation with the Major about the accident on the Cape."

Kennedy sat up and started. "There's not much to tell. He asked me how I knew about the accident, but didn't know about the moon landing. Told him I called home every Sunday night, my wife and my parents. My wife told me my parents weren't home, and I should call Bobbie. When I called Bobbie's house, I learned about the accident. Then he asked if I talked to Bobbie, and I said yes. Then he got antsy or nervous, I don't know what you'd call it, but he signed my papers and asked me to call him if he could be of further help. That's when he gave me his card."

"That's it?"

"Yeah, why?"

I passed the paper to him and told him to read the story on page three. As he read, I watched as his eyes widened. After only about thirty seconds, he put the paper aside. "Holy shit. The accident I was talking about was my sister in Barnstable, but he was talking about the senator and Chappaquiddick."

"And," I added, "he thinks Bobbie was Bobby. Sounds the same when you say it."

Kennedy laughed, "That explains the royal treatment."

"How so?"

"Once I got back from Post Personnel, they treated me like one of the staff. They invited me to have meals at the cadre table, talking and joking like I was one of them. When I went to the post tailor, they made sure the guys there fit me proper. I didn't have to walk anywhere.

There was always someone to drive me and wait to take me back."

Kennedy broke the silence between us after a long minute. "The guy's a moron. He'll probably tell the story of how he helped a Kennedy. Would love to be there when someone points out Bobby was killed last year."

Back in my seat, I thought of the circumstances around our assignments; mine, a missed announcement and fibbing my way back into the interview; his, a misunderstanding about two accidents on Cape Cod. The rigorous selection process for the Pentagon was proving to be something less than rigorous.

The urban sprawl of Philadelphia gave way to the lush green of northern Delaware, replaced all too soon with urban blight as the bus slowed and exited the interstate in Wilmington, Delaware. The big oval DuPont sign on a building off to the east disappeared, hidden behind the tenements, radiator repair, and pawn shops that seem to surround bus stations in large cities. The bus stopped at the curb. Wilmington must not be big enough to have a terminal, just a bus stop. One passenger got off and several got on. The new passengers moved to the back in search of a seat.

The solitary seat I enjoyed from Philadelphia became a thing of the past. A tattered man took the aisle seat to my right after placing an equally tattered suitcase in the overhead rack. He stuffed a beat-up paper bag under the seat in front of him. The paper bag was a miracle of survival. It was so worn it seemed to have lost all sense of being made of paper. The bag didn't have any creases or folds, and it collapsed on itself to mold to the place underneath the seat and to the stuff inside. So soft, it looked as close to leather as a paper bag could get. Unlike the bag, the man had all the wrinkles of being well worn, and made lots of creaking sounds as he tried to make himself comfortable. Had he carried the paper bag so long it robbed him of his pliability and quiet, for itself?

The man appeared down on his luck. I couldn't guess his age. His dirty blond hair curled around his ears and collar, not in a fashionable style, but in a "need a haircut" manner, and his clothes were dirty and threadbare. Is he wearing them because his good clothes are in his luggage, or was this the best he owned? I suspected the latter. His hands were small, and so dirty the fine blond hairs on the back of them looked like they had black roots. These dirty fingers dug a cigarette out of the pack he extracted from his T-shirt pocket. His hands trembled as he lit the smoke. He expelled the smoke as he looked out the window to his left, which meant he blew the smoke directly in my face.

I wanted this time to think about my future and how lucky I was to be safe in the Pentagon, instead of not-safe in the jungles of Vietnam. But he wanted to talk and blow smoke in my face. He wanted to know where I was going and where I was from. When I didn't answer, he wanted me to know he was a vet, too. He wanted me to know he was on his way south to visit his son in North Carolina.

I'm a smoker, lighting them up just about anywhere and anytime, but not on the bus. Something about smoking on a bus made me queasy. Maybe the motion of the bus, or the smell of diesel fuel exhaust, or maybe the combination of it all made me avoid it. Today, there was another component to the motion and the diesel fumes. Today, the unwashed, sour stench of this man brought things to a boil in my stomach—which was empty. Hours had passed since breakfast at Fort Dix, and I'd only had one bite of a cold hot dog that became a mushy ball on the roof of my mouth. That thought was enough to produce a gag as he lit up his second cigarette, just two minutes after grinding out the first one in the tiny ash tray in the arm rest of his seat.

I spoke to him for the first and only time. "If you promise not to blow any more smoke in my face, I promise not to throw up on you." I punctuated my sentence not with an exclamation point or a period. I punctuated it with another gag.

It proved enough to get him to reconsider where he wanted to sit and smoke. He grabbed his soft leather-like paper bag and moved further back in the bus, leaving me with an empty seat beside me, cleaner air around me, and my thoughts.

The bus moved steadily south. The rhythm of the big tires thumping over the spaces in the poured concrete seduced me and I fell asleep, waking only twice. Only briefly when the bus stopped in Baltimore, and completely when it slowed for our arrival in Washington.

CHAPTER 3

"Fort Myer, the one without the 'e' and without the sand"

Ah, Fort Myer. The name conjured up visions of a resort in Florida, on the glorious beaches of the same name, relaxing around the pool, watching the sun set from the veranda outside of your room. But we weren't there yet. We were at the bus station in Washington.

Was it hot? Was it dirty? Was it smelly? Yes, it was like the bus station in Philadelphia, but hotter, dirtier, and smellier. It was also noisier. Kids crying because they didn't want to go to summer camp. Parents pleading because they needed their kids to get on the bus and go to summer camp, if only so they could leave this smelly cesspool and get back to their air-conditioned homes.

Outside, the blazing sun's hot rays reflected off the dirty chrome bumpers, dusty windshields, and faded paint of the dozens of cars, taxis, and buses that converged on this transportation hub. These reflected rays—together with the residual heat from a week of hot and humid summer yesterdays trapped in the buildings and the streets—reminded me of one of those French paintings. Paintings where all the colors and shapes are wavy, blend into each other, and you have to tilt your head to figure out what you're were looking at.

There weren't any cars or drivers waiting to take us to Fort Myer. I thought at the very least they'd have one of those olive-green, mini-school-buses to take us there, wherever "there" was.

My orders had an idiot line on them. An idiot line is a section that tells idiots what to do in case they get lost. If you're just starting your trip, it instructs you to call the place you left. If you're almost there, it gives you another telephone number, at the place you're going. The Army won't let you go anywhere after you enter Basic Training until you have some basic skills, like brushing your teeth, wiping your ass, shining your shoes, and reading a map, all the Army way. The earliest they'll let you go anywhere is five weeks into your training. They figure, they can teach you in five weeks a whole lot about survival in this world you've been living in for twenty years. With this training, you're then skilled enough to go home and visit your momma. And if they're wrong, and you can't find your way, there's always the idiot line.

The Army put me in charge of moving us from Fort Dix to the Resort and Spa at Fort Myer. So, I got my orders out and called the "Destination Number" listed on the idiot line of the orders. I got someone who identified himself as CQ.

CQ is another one of the Army abbreviations we had to learn. It stood for Charge of Quarters. CQ was one of the more interesting abbreviations because it could be a person, a place, or an activity. It wasn't a permanent rank or place. A soldier got the assignment of being CQ for a shift, sometimes a day, or more likely overnight. The place for CQ was usually the company clerk's desk, or at night it might be a desk in the supply room. Even though it could be three things: If someone was CQ, he was CQ doing CQ in the CQ. Three different things, all the same, the Army way.

"Hey CQ, this is Private West," I announced.

No comment from CQ.

"Private Jonathan West, with the soldiers from Fort Dix."

No comment from CQ.

"So, we're ready to go. Right over to Fort Myer. We can't find the Army bus that's supposed to take us over to Fort Myer."

CQ finally talked. "How did you get this fuckin' number, soldier?" he asked. It was more of a bark with a snarl attached to it.

I had the answer straight away. "The number is listed right here in paragraph 3.2.1 of our orders, dated 30 July 1969," I responded.

"Soldier, that section is the idiot section of the orders. It's for emergencies or if you're lost. It's not to have me make travel arrangements. Do you have an emergency, Private?" he snarled.

"No, CQ," I responded.

"Are you lost? Because if you're lost, you had better turn around and go back to Fort Dix and take that sorry ass group you have with you. We don't have room for the lost or losers here. Do you understand, Private?" he barked.

I was getting an image of CQ that wasn't very pretty.

"No, CQ, we're not lost. We just wanted you to know we're on our way. Didn't want to miss our transportation to the fort if it was on its way. We'll grab a cab—"

Before I could finish, he interrupted me. I could feel CQ sit straighter in his chair as he asked, "Is this Kennedy's group?"

"Yes, Kennedy's in the group, and I'm sure he's okay with taking a cab."

"Just tell him we're sorry and a little short-handed, what with Eisenhower's funeral and all."

"Sure."

Dexter W. and Kennedy were disappointed there wasn't transportation for us, but understood the predicament the CQ was in.

"Shortages of everything is part of the war effort. That's why we've been called to the Pentagon," responded Kennedy.

I grunted agreement.

"Eisenhower died in March," said Dexter W.

"Shut the fuck up," I said.

"I read it in the newspaper," he responded.

"No shit. CQ must have it confused with something else."

While I was on the phone with CQ, Dexter W. and Kennedy struck up a conversation with some taxi drivers around the bus station. When they found out we wanted to go "all the way to Fort Myer, way over in Virginia," they didn't want to take us.

"That's out of state, you know. If traffic's bad, we sometimes hafta go through Maryland first. Then that be two states to get you there. And it's rush hour, and with all your shit"—he said, pointing to the duffel bags with the hangers on the outside—"we'll hafta take two cabs."

One of the ladies of the night working the day shift interceded. She took one of the drivers aside and talked to him. She came back and told us she had arranged a special price with her friend of twenty dollars a person, plus tip. Not knowing where Fort Myer was, I asked if that was a reasonable price. She agreed it was very reasonable, considering it was out of state, maybe two states.

"Sweet thing, they usually get thirty dollars from you guys, but we worked out a special deal, just for you."

I couldn't believe how sweet she was. She had done us a favor getting our ride to Fort Myer. I gave her a two-dollar tip. She was thrilled.

Off we went, duffle bags stashed in the trunk. The girls were lined up, giggling and waving and having a grand old time as they sent us on our way.

We settled in for the long ride out to Fort Myer. Traffic was surprisingly light, we were told. Before we knew it, we were taking in the sights of Washington. As a kid, I had been to Washington several times with my parents. I remembered Washington, D.C., in August as being very hot. How hot? We had a '55 Ford station wagon with rear windows you could open the top half, to let the breeze blow through. It was so hot that instead of a breeze, it felt like an oven, cooling down only when the car stopped. Washington, D.C., in July was even hotter.

The driver took us along the Mall, with the White House on the right and the Washington Monument on our left. Within minutes we passed the Lincoln Memorial and crossed the Potomac on the Memorial Bridge. A quick right after the bridge, and we were on the solemn grounds of Arlington National Cemetery.

"It was nice of you to take the scenic route," said Kennedy. He had been trying to engage the driver in conversation since leaving the station. Wanting to know how far it was, how long it would take us to get there.

"Won't be long. We'll be there before you know it." The driver sounded like every bored father who has ever taken his family on a drive.

"Leave him alone," I said, "You sound like a fucking five-year-old . .

23

. When are we going to get there, can I have a drink, I have to pee, are we there yet, are we there yet?"

"Okay. We here," said the cabbie as he pulled to a stop.

I laughed. It was the first time he talked to us. Listening to Kennedy whine and me trying to shut him up, he'd finally decided to talk. After all, we were going to spend a bit more time in his cab. We'd only been driving for about ten minutes. Conversation would probably help pass the time for him on the long ride all the way to Fort Myer in Virginia.

"See, Jim, even a professional driver who hears all kinds of stupid whining day in and day out has given up. My father used to say that to my little sister. She'd say 'when,' and he'd say 'just around the corner.' He could 'just around the corner' her for thirty miles."

"I ain't fuckin' with you, man. We here. This be Fort Myer. We is all the way into Virginia! Where do you want me to drop you and your shit off?"

"I can't believe it. Twenty fucking dollars for a twelve-minute cab ride." Dexter W. was pissed. "Should have taken the bus." For his twenty bucks he'd extract twenty dollars' worth of "I was right, and you fuckers were wrong" righteousness. Everything we did now would be measured in multiples of this cab ride.

"Come on, you dickheads, we're here," I said, getting out of the cab.

We collected our duffle bags and hangers from the trunk. The driver stood waiting. I said thank you, and I meant it.

He said, "Tip."

Dexter W. said, "Go fuck yourself."

"I overcharged you anyway, so go fuck yourself, yourself."

CHAPTER 4

"Meeting CQ in the CQ"

We had arrived at the same post where General Patton had played polo before he became a movie star. The Army selected us from thousands of soldiers to come to the Pentagon and make our contribution to end the war in Vietnam. Cocksure, we strode into the CQ to announce we were ready to go to work.

The CQ reminded us we were still in "this fuckin' man's fuckin' Army" even though we would work at DA, and we had better take our hats off in his house.

The Army must have a requirement that anyone who greets new soldiers to the post must act like a jerk, be they recruits to Fort Dix or "soldiers" to Fort Myer. The guy who called himself the CQ was a pimply faced Corporal still in his teens. Mother's milk was on his breath.

"Who the fuck is in charge of this gang?" barked Corporal CQ.

"I am," I said.

"Not anymore, dickhead! Look at these fuckers standing around like they're back in neighborhood. I watched these dickheads getting out of the cab, walking down here like they had all fuckin' day. This here is the stractest fort in the world. We gots the Third Herd here.

All the fuckin' Generals in the world livin' here. We even got horses, and a cemetery. And now we got you dickheads." CQ shook his head, walking back and forth. His ranting had us standing up straight, not at attention, but standing straight.

"I beg your pardon," I said. "Sergeant Cowlings put me in charge of getting us from Fort Dix to Fort Myer—"

"Dickhead, I didn"t ask for your opinion. But as long as you opened your pie hole, let me put you straight. Where are your Sergeant Cowlings now? Why, he's in Fort Dix. And who's here? It's me."

"But—"

"BUT?" The Corporal moved in front of me. "You gotta be a college boy. All you fuckin' college boys wanna debate alla time. Let me tell you this, college boy. You was under orders to get you and these peckerheads here from Fort Dix. They put you in charge, right?"

"Yes."

"Yes what, dickhead?"

"Yes, Corporal."

"So, Sergeant Cowshit put Private Dickhead in charge of bringing these peckerwoods all the way down from Fort Dix. Well, Private Dickhead, you got them here. New place, new rules. I'm in charge now, not Sergeant Cowshit. You!" he said, looking at Dexter W. "I'm gonna put you in charge of taking Private Dickhead and the rest of the Fort Dix peckerwoods over to chow. Feed them and get them back here, while I figure out where I'm going to billet you tonight."

"Yes sir, Corporal. I'd be proud to take command."

"Shut the fuck up. The only thing worse than a college boy is a smart-ass college boy. Get them to the mess. Be back here at 1800 hours. For you dickheads, that's six p.m. For you smart-ass college dickheads, that's when the big hand is on the twelve and the little hand is on the six. And take this other useless college boy with you," he said, looking to another Private who emerged from a door in the back of the office.

After ten weeks at Fort Dix, any sentence having both chow and mess in it caused me to picture a red-and-white-checkered Purina bag with a picture of a white furry tongue looking at me.

Outside, the newcomer introduced himself to us with a big southern accent and an even bigger smile. "Leslie Michaels, but I go by Les. You pronounce it 'Les,' like in 'less', not 'Les' like in lesbian." As we walked out of the basement of the Consolidated Barracks to the Consolidated Mess Hall, Les told us he had arrived earlier in the day from Fort Polk, Louisiana.

Barracks and mess were "Consolidated" at Fort Myer, because the facilities were shared by the Army, the Navy, the Air Force, and WACS who worked in the Pentagon. A sidewalk took us across a great expanse of well-maintained grounds, with grass and bushes neatly trimmed, to the modern-looking Consolidated Mess Hall.

We could see the older barracks of Fort Myer in the background. These buildings were the home of the famous Third Infantry, or Pershing's Own, the soldiers who guard the Tomb of the Unknown Soldier and escort the caissons in the funerals at Arlington Cemetery. The Third Herd, as they were known, accompanied President Kennedy's body on that long walk just six years ago.

The mess hall was air-conditioned, a welcomed relief from the heat. There were no crisp white linens on the table. Didn't appear to be any waiters to complement the cafeteria. The inside was bigger and brighter, but looked pretty much like the one we had left at Fort Dix, with one exception: women! There were a lot more females here in one place than any of us had seen in more than three months. The females were the enlisted Navy WAVES, and Army WACS, and whatever they called the Air Force women.

Leslie was in heaven. "I smell pussy," he said. I can't say I did. The room smelled like a cross between the high school cafeteria and the old mess hall at Fort Dix, but without the sweat.

We showed our ID cards and our orders and got in line, grabbing trays and eating utensils from under the "Eisenhower sign."

I saw the sign when I first arrived at Fort Dix, about 1 a.m. on a bus from New York City, with over thirty other new Army inductees. After they made us do some paperwork, they rushed us off to the mess hall. We thought it was nice of them to keep the mess hall open so late, until we learned it was always open for newbie arrivals at this time. Feeding new arrivals in the middle of the night would not mess up the routine of the permanent soldiers. This was the first place I saw the Eisenhower sign: "Take all you want. Eat all you take." It was called the Eisenhower sign, because the saying was supposedly attributed to Ike when he was doing the D-Day thing. Nice sentiment—inviting, but cautious. Problem was, the soldiers working the mess hall at night didn't like working at night, and didn't like us because we were the ones responsible for them having to work. So, they made our collective lives miserable.

It wouldn't be much of a stretch of the imagination to think I might get eggs, pancakes, and other breakfast stuff at two or three o'clock in the morning. Think again. Instead, we got what the soldiers the night before got, and didn't finish. We got pork chops, hash browns, corn, and beans. And by beans, I mean the stuff that failed the Campbell's pork-and-beans quality check. And by "got," I mean we were served by the misfits working the graveyard shift. Just like every Army film ever made, we each took our tray and went down the line. We got two greasy pork chops, a big gob of cold hash browns, kernel corn, and a ladle of beans over everything. At the end of the line, I grabbed a bowl of green Jell-o, thinking it was the only thing appealing. So did a lot of the others.

When we sat down, all hell broke loose. People shouted at us to eat faster. "You mother fuckers got four minutes to chow down." I focused on the Jell-o. A mistake. I took my time savoring it despite the hellstorm brewing around me.

"Time's up!"

"Where do you think you're goin', mother fucker? Finish that chow!" a mouth roared. But I'm not hungry, thanks; the Jell-o was enough. Then one of the mouths pointed to the sign, the Eisenhower sign. Then came my first order: Finish that chow. I tried, but I couldn't. This led to my second order. "Get down and give me twenty." Twenty what? "Twenty-five push-ups. You must be a college boy, axing so many questions. Are you a college boy?" Yes. "Then you can get down and give me thirty."

I gave him thirty, while the others gave anything from ten to twenty. When we finished our exercise, we made the mistake of thinking we were done. "Now you must be hongreee after all your exercise. Sit down and finish your chow."

We did this as a three-act play: eat, push-ups, eat, until they grew weary of playing with us. The mouths let us empty our trays in the garbage, but assured us we were marked for disobeying a direct order from General Eisenhower.

I took a tray from under the Eisenhower sign and set out for the serving line. I held back, watching the line in front of me. Surprise, surprise, surprise! No greasy pork chops. No one slopping gobs of food on the tray, but rather someone asking politely: What you would like, and how much? I don't remember what I had that first night at Fort Myer, only that it wasn't pork chops, it wasn't beans, and it wasn't too much.

The four of us ate as a group, in relative silence. Tired, yes, but more likely we still had residual purposeful resolve from Fort Dix—get the food in your mouth before someone told us to clear out, with thirty seconds' notice. I watched those around us. Some went up for seconds. Some were finishing their meal and going for a cup of coffee. Others got coffees for their table, enjoying conversation, coffee, and a smoke after their meal.

This place might be civilized.

Lulled into thinking we had become elevated above the crap we were used to, we strolled back to CQ to get our rooms for the night. CQ would be happy to see us!

"Where the fuck have you been?" CQ didn't seem happy to see us.

"Grab your gear and follow me."

Grabbing our duffle bags, we followed him down a hallway to a closet, where he issued us sheets and a pillow case. "Stow them away and follow me," he said as he went back down the hallway we had just left to the room we had just left. We must have looked like we needed the exercise of carrying our stuff down and back, just to pick up some overnight linen. More likely, an early indicator that life in the real Army was just as much a waste of time and energy as life in Basic Training.

"Take all your shit and go out in front of this barracks. Get on the bus to South Post. Report to the CQ at South Post barracks for your housing assignments."

That was it. No kiss goodbye. No "stay in touch." No "sleep well." Just get on the bus, and get out of my sight.

We trudged up the ramp to the parking lot at the back of the Consolidated Barracks. This was the first time I could take it all in. This was no ordinary Army barracks. More the size of a college dormitory—bigger than most college dorms. We walked around the building, looking for the bus stop, sweating in the early evening heat from the mild exertion. CQ was waiting for us! There must have been a shortcut through the building.

"Who's the fuckin' idiot in charge of this circle jerk?"

"I am, CQ," Dexter W. said.

"Orders" is all the CQ said as he handed him the packet left behind. Well, not really left behind. I gave it to him when we reported, and he didn't give it back until just now. "Fuckin' idiots," he added as he walked back in the entrance to take the shortcut to his office.

"I think he was talking about you guys. If he had meant me, he would have used the singular—idiot," I teased.

Dexter W. took charge. "OK, you fucking idiots. Find me the bus to South Post and our next adventure with our next CQ."

"You're in charge, you find the fucking bus."

"I'm delegating."

"Ain't no such thing as delegating in this man's Army."

The debate ended when a bus appeared with a South Post banner. We grabbed our gear and got on. Then we settled in for what was truly one of those memorable experiences you get only once in your life.

CHAPTER 5

"Off To See CQ Two"

The bus left the modern area of the Consolidated Barracks and Mess Hall and passed through the area dedicated to the Third Infantry. These were the barracks of yesteryear movies, two-story red brick, with a porch running the length of both floors. Everything neat as a pin. The soldiers we could see worked with a purpose. No one was sitting around talking, having a smoke, or just relaxing. Quite a contrast to the almost college atmosphere in the modern Consolidated area. We were less than a city block from our start at the Consolidated Barracks, and it was like we had journeyed a hundred years into the past.

The Third Infantry was referred to as the Third Herd, and they were "super strac." "Strac" was a term meaning a soldier took time and effort to get themselves and their gear together. A little more time to shine our boots, a little more time to shave, and a little more time to iron our uniform. "Super strac" meant you could shave with the crease you ironed into your trousers, using the shine on your boots as a mirror. The Third Herd never walked anywhere. They marched or ran with purpose.

Another city block brought us to downtown Fort Myer. Here, they provided all the trappings of a small town: movie theater, post office, NCO club, and the officers' club. Again, all red-brick buildings from another era, emphasized by the smell of manure from the stables just off the main street—stables that housed the horses used in the funeral services at Arlington National Cemetery. A silence fell over our small group as we experienced the same awe, the same reflection to that terrible week in November, just a few years ago.

Then the mansions started in the section of Fort Myer I would always refer to as "Generals' Drive." Here sat the homes of the Generals assigned to work in the Pentagon. Immaculate red brick with brilliant white trim. No toys in the yard. The lawns and shrubbery trimmed to perfection, not a blade of grass not standing at attention. Big stately trees provided ample shade for these homes. On the edge of this little community was a non-denominational chapel, used by all faiths for sabbath services and for the funeral services at Arlington National Cemetery.

Down the hill, and around the corner, and there it was: Arlington National Cemetery. The cemetery separated North and South Post of Fort Myer—or to be more precise, Fort Myer guarded the cemetery with the North and South Post. Off in the distance was Washington, D.C.; the view familiar from the Kennedy funeral procession. We were seeing it from an angle of about forty-five degrees from that infamous television shot, but it was still recognizable and truly remarkable. The long expanse of precisely aligned graves led down to the ceremonial entrance of the cemetery. There, the bridge to Washington, and the Lincoln Memorial at the end of the bridge, all stood in brilliant white marble, shining in the evening sun, surrounded by the blue of the river and the green of the lawns in the parks.

I was in awe. What a sight!

"I think we're heading north," Dexter W. said, disturbing the reverent silence. "If we're heading north, how come this place we're going is called South Post? Shouldn't it be North Post, and the one we just left be South Post? What I mean is, all of you remember the Kennedy funeral. Do you remember Walter Cronkite being the commentator?"

"Yeah, so what?" I answered.

"Well, I remember him saying we were watching the procession head south to Arlington Cemetery. If that's the case, this can't be North Post."

Les jumped on this. "Well, if Walter Cronkite said so, then it's gotta be the truth. Where I come from, if we was sitting down for supper and Walter came on the news and said, 'Good morning,' my daddy would get up and go to work and my momma would start tellin' us to get ready for school."

I looked at Les, who had a big smile, then to Dexter W., who had a peculiar look. He had just gotten support from Les, but it looked like he was unsure of its validity or sincerity.

Wrong thing to say to Dexter. "Driver. Do you know why they call this South Post when it is really north of North Post?"

The driver, a civilian but probably a government employee, answered, "What?"

Dexter W. repeated his question. The driver responded, bored, uninterested, and unimpressed with the question. "It's a name, son. Just a name."

"But it's wrong. South Post should be North Post, and vice versa."

Kennedy took out a notebook and started writing. He had told us he was going to write a novel about his experience in the Army that would be a combination of From Here to Eternity, Catch-22, and the movie M*A*S*H. Talk about ridiculous! Who would waste their time writing such a thing? Who would waste their time reading such a thing?

I jumped in, "I don't give a shit if South Post is North Post, or if North Post is the South Pole, or if we're going north or south. I just know we ain't going to South Vietnam. So, shut up before your stupid questions screw this thing up. We got picked because we're supposed to be smart, and it doesn't sound smart to be asking stupid questions."

Our driver brought the bus to a halt beside a raggedy-assed building, and announced we were at South Post barracks. What a shit-hole this place was! It looked like it was built pre-World War II, and hadn't been painted since the Korean War. We rolled off the bus with a lot less awe than we had when we were driving past the Generals' mansions.

"Bye, sonny," said the bus driver as Dexter W. passed. "Welcome to the South Post in the north. Ha!" he added, laughing at his own joke.

We stood, not knowing which way to enter the building. One end was guarded by a garbage dumpster, and the other end was a screened-in porch. We opted for the porch entrance. A door from the porch led into the building. Les volunteered to stay with our gear on the porch rather than have all of us drag it through the building. It would be just our luck we'd haul our shit all over the building, only to find our sleeping area was just inside the door. Les's idea was good. We were going to start out-thinking the Army.

One thing was certain as we entered the building: It wasn't air-conditioned, like the Consolidated Barracks. It was hot. A center hall ran the length of the building, with cubicles the size of prison cells lining both sides. While affording some degree of privacy, the walls defining the cubicles only went up six feet and probably offered little protection from sound. The good news: The open ceiling space aided in air circulation. At the end of the hallway we found someone we could ask about CQ. We were close; CQ was at the back of the building, on the second floor. We found him at the top of the stairs.

This CQ seemed a little more laid back than the one on North Post. He was in a T-shirt, had his bare feet up on the desk, and was doing absolutely nothing. He wasn't reading, or sleeping, or watching television, or listening to the radio, or talking with anyone, or smoking. Just sitting there, eyes open, with his bootless feet up on his desk.

"CQ. We're the guys from Fort Dix. We were up at North Post. CQ sent us down here for room assignments." Short and to the point, I thought.

"Jesus H. Christ, it's hot," was all he said. I didn't know if he wanted a response, or to be left alone.

"Fuckin' A," added Kennedy. "I need a shit, shower, shorts, sandals, and a six pack of 'gansett."

"You from New England, rookie?"

CQ was the first NCO I had heard speak two consecutive sentences without having to use "fuck" since that first night/morning arrival at Fort Dix. He was also the first NCO not to call us dickhead, or dude, or shit-head. Rookie was a civilian term. Tom Corbett to earth: "We have found signs of friendly life on this forbidden planet."

"Worcester. Grew up in New York, but went to Holy Cross."

"No shit. Knew you were a local, sort of. Nobody but us locals will drink that 'gansett cat piss." CQ extended his hand across the desk. "Jimmy Dimao, Providence High, then the University of Fort Dix, with graduate work at the University of the 'Nam. Getting the college boys, now?" A question that didn't need an answer.

"Well gentlemen," he continued, "I've been expecting you. You've met the HQ Company CQ up at North Post. He either has the rag on today, or you pissed him off. Probably has the rag on, because he has the rag on every day. Jody got his girl, is the word."

Jody is the name given to guy back home who's stealing your girlfriend or wife.

"Anyway, he said he wanted me to give you real special billets while you're waiting for room up on the hill. Problem is, I got nothing but special billets—if you know what I mean. We are full. If you were Joseph and Mary, I'd have to tell you I got no room in the inn."

"I'm not going back up there and sleep in the stables!" declared Dexter W., not to anyone, just more of his thinking out loud. The Army did someone a big favor by not sending him to Vietnam. His thinking out loud would get someone killed in Vietnam.

"Sorry there…" CQ was waiting for a name.

"Dexter W. Green."

"Sorry, but I don't even have room in the stables," said nice CQ as he got up. He went to a closet door, opened it, and tossed pillows to us.

"They told me they issued you blankets, sheets, and a pillow case up at Consolidated Barracks. Here's a pillow to go with that. We don't issue blankets in the summer in Washington. Don't know why you got them. Where's the other guy and your gear?"

"We left him watching our gear on the front porch. Saw some guys walking through there, and we didn't want to take a chance."

"Probably not necessary to leave a guard. Everybody stationed here considers themselves lucky. No one is going to take a chance on an Article 15 and getting shipped to Vietnam over any shit you guys might leave around. And by the way, you didn't leave him on the front porch. Here at South Post, the correct military term for that area is 'your bedroom.'"

Kennedy and I looked at him, then at each other, then back at him. "Our bedroom?"

"Sorry guys, no room at the inn, like I said. May be a bit public, but, unlike the stables, there's no horse shit."

We followed him down the stairs and back through the hallway, eyeing the cubicles that now looked like a room at the Ritz. He pointed out the latrine on the first floor as we passed it, but called it the "bathroom." Les was where we left him.

"Hi, fellas. Nobody touched any of our stuff. Real friendly bunch of guys come through here. They smiled and said hi. Some didn't say hi, but they all smiled. This here porch is a nice friendly place."

Nice CQ turned to me and said, "You're in charge. Tell him what the official military term for this porch is."

"Leslie—" I started.

"Less-lee," he corrected.

"Less-lee. The correct military term for this area is 'our bedroom.'"

A few seconds of thought, then the ever-optimistic Less-lee said, "Then it'll be just like staying at Gramma's at the lake. I sleep on the porch. When I need to take a leak in the middle of the night, I just go to the railing and pee on a bush."

"No peeing on the bush. That's an Article 15. Tell him why we don't like Article 15s here, West." Nice CQ must have a lot of confidence in me as a leader, because he was having me do all the explaining.

"Article 15 means no Pentagon. Article 15 means Vietnam."

"Okay, gentlemen, that's it for the night. HQ CQ wants you back up at North Post, same place, tomorrow morning at 0800."

"We get to sleep in. I thought the Army started everything at 0500," I offered.

"The Pentagon starts at 0800. If it's good enough for them, it's good enough for the rest of us. Mess hall is across the street for breakfast. Bring your orders to get served. If you already ate, you can go in now and get coffee or Coke. Sorry, no 'gansett," he said, smiling at Kennedy.

"Corporal," I needed his attention. "We're all pretty tired. You know, the excitement of coming to work in the Pentagon, and being on the road all day. Any chance someone can come by around 0600 and make sure we're up?"

"You can be sure someone will be by. You won't oversleep," he said with a wry smile.

"Pick a spot and settle in. No bitchin'! There's guys in 'Nam sleeping in muddy holes in the ground tonight," I said, trying to convey a sense of duty for them and authority for me.

We began to settle in, each getting a corner.

As I rolled out my sheets, Kennedy said, "There ain't no 'gansett, but that don't mean I don't get my shit and a shower. Then I'm going over to that mess hall and have a coffee, a piece of pie, and a smoke. Do you get that, 'rookies'? I got free time in this man's Army for the first time ever, and I'm going to act like I own it."

He was right. This was the first time in over ten weeks we didn't have to be someplace, or have someone telling us to do something, or pretend to be doing something.

"Me, too," echoed two times.

CHAPTER 6

"On Our Own"

In the heat of a late July evening, we could hear Christmas music, The Twelve Days of Christmas, playing in the distance, maybe at the back of the building on the second floor. Someone trying to beat the heat without air-conditioning. Les sang along, making up his own lyrics.

"Three soldiers shitting…"

Joining in, I added, "Three soldiers showering…"

Kennedy added, "Three soldiers settling for something sweet."

"'Stead of suds," as a chorus, continuing our friendship bonding alliteration.

"'Stead of pussy," corrected Les.

"''Stead of pussy,' said the stranger on the steps," sang the stranger on the steps, who had just walked through our bedroom/his front porch on his way out.

"Son of a bitch. Sure do like this place. These fellas is just real friendly." Les was all smiles.

Here we were. Our station in life was improving. Last night we were sixteen men to a room in Fort Dix, and tonight we were four soldiers in a room with cross ventilation, easy access to transportation, and just a short walk to a dining facility. I think we broke even on distance we had to travel to take a dump.

The irony of the term private struck me. I had been a Private now for almost twelve weeks, and the last word I would use to describe my experience would be "private." "Public" was more appropriate. In contrast, the bus ride from North Post had exposed me to the Generals' houses, and they were anything but "general"; more like privilege, or private. So, the terms were dreamt up, no doubt, by the same oxygen-deprived moron who coined the term "Military Intelligence." Is oxymoron a contraction for oxygen-deprived moron?

So, we four soldiers stopped singing, shat, showered, and went in search of something sweet— down the steps, across the street to the mess hall, where the singing stranger on the steps had shortly gone before.

The South Post Mess Hall didn't look like a mess hall—at least, it didn't look like any of the three I had seen in the last twelve weeks. Granted, my personal reference points were limited; but I'm sure memories of those I've seen in movies or on television would have rushed to the front of that nerve gob behind my eyes and said, "Hey, this looks the mess hall in From Here to Eternity." But all the nerve gob was sensing was this mess hall was like a 1940s diner, without the counter and stools.

The entrance to the air-conditioned mess hall was at the back of the building, into a hallway separated from the eating area by a half-wall that had flowers in vases sitting on top. Yeah, flowers in vases, in an Army mess hall. At the end of the walking area, a soldier sat at a table. We showed him our orders, and he allowed us to pass into "the best-kept secret in the whole fuckin' U.S. Army."

It had to be the best mess hall in the world, in any army.

The room wasn't fancy, just sort of comfy. We all felt it that first time. The dark green tile that was the hallway continued into the rest of the room, complemented with tans and blacks that worked well together. The eating area had six ceiling fans, turning slowly. It sounded different. The buzz of conversation replaced the bellowing of cadre screaming. But even compared to the civility of the mess hall on North Post, there was a difference. There was no clank of serving utensils on metal pots or plates. Tables were set up to seat four or eight. No checkered table clothes or Chianti bottles with half-burned candles in them, but it was the kind of place it could have been pulled off. The biggest difference was the smell. Not the smell of a mess hall; not the smell of a high school cafeteria; but the smell of a restaurant.

Smell is one of the senses evoking the most profound memories, be it the smell of fresh-cut grass, the perfume of your first girlfriend, or the ocean. All pleasant reminders. Some smells can evoke both pleasure and fear, like fire. Is it a friendly campfire, or a forest fire? In early man, and perhaps in the creatures we evolved from, smell was a critical sense for survival. While evolution diminished the sensitivity of our olfactory lobes, it resisted reducing the imprinting of these smells in our memory.

Enough of the ambience; we were about to get our first taste of why South Post was the best.

After getting trays and silverware, we started across the front of the room. The building was about eighty feet wide with the entire front of the building dedicated to serving food. The sign at the entrance listed the breakfast, lunch, and dinner hours. The "other" hours were in effect now, and described as a "limited grill." Makes sense. In the military, there are folks working 24/7, and they must be fed. The South Post Mess Hall is the closest mess hall to the Pentagon, within walking distance of the building, a distance defined as less than the time it takes to smoke a cigarette. All the soldiers, sailors, and airman who work something other than the day shift must be able to get a wholesome meal at any hour of the day or night.

As I started down the serving line, it was difficult to understand the "limited" part of the "limited grill." It seemed all the containers on the steam table were being used and full. Lots of vegetables, two kinds of potatoes, pasta, chicken, and fish. On the other side wall, opposite the hallway, was a salad bar and dessert station. I saw what I came for, coffee and dessert. As I passed the steam table, I saw the grill area. I didn't stop, just slowed down to see what was available.

"How can I help you?" asked a cook behind the counter. Smartly dressed in his white cook's outfit, complete with cook's hat, he stood, not at attention, but attentive, waiting for my response.

"Whatcha got?" I asked.

"Just about anything you could want."

"How about a steak, medium rare?"

"Monday night and Friday nights only. Be glad to help you then. If you be new here, you be best advised to know Monday night is better— less of a wait."

"You came up short on the 'just about anything you want' offer," I said.

"The 'just about' just about covers that," he responded.

I'm going to have to stay on my toes here, I thought; even the cooks have all the answers.

"I guess I'll have to come back for one of your steaks then. For now, how about a cheeseburger, the way my Uncle Chuck used to make them?'

"Tell me how Uncle Chuck used to make them, and you'll think he was standing here next to me making it hisself."

"First off, medium rare, but the outside is charred, with white Kraft American cheese. The cheese has to be dripping off the burger, but not touching the bun. Put the bun on the grill, but just to warm it, not to make it crunchy. The burger with the cheese sits on the bun and the juices from the burger just seep into the bun, not too far. And all of this on a charcoal grill."

"Can do," he smiled, "One regular cheese burger, just like I always make."

"Great," was all I could manage. This guy was going to give me last night's left-over meat loaf on a bun, or he was for real. I hoped he was for real.

The big, muscular cook turned his back and got to work on my order. I couldn't see what he was doing. His size blocked the entire cooking surface. I could hear the clatter of the metal spatula on the metal grill. I could hear the sizzle of meat, and I could smell the initial sear of the meat. He looked over his shoulder, not able to make a full turn to face me, his huge neck restricting the movement, and said, "Some of those boys from California want that Monterey Jack cheese on their burger. They can have it, but it ain't no Uncle Chuck burger. An Uncle Chuck burger is an American burger, and American burgers get American cheese.

"Some officers want Swiss cheese on their burger. They can have it, but it ain't no Uncle Chuck burger. An Uncle Chuck burger is an American burger, and American burgers get American cheese."

I didn't know if he was talking to himself or me. He was still neck-restricted, but I think the "Uncle Chuck" stuff meant he was talking to me.

"Some of them boys from Philadelphia want Cheez Whiz on their burger. Say they have a sandwich up there that's grilled sirloin tips with Cheez Whiz. They can have it with Cheez Whiz...." The big man stopped talking, shifted his feet so he could look over his shoulder at me. He raised his eyebrows and nodded at me and said, "But...?"

After a few blank seconds I got it, and continued, "But it ain't no Uncle Chuck burger, 'cuz an Uncle Chuck burger is an American burger, and deserves American cheese."

"Close enough. You'll do better the next time." He plated the burger and brought it to my tray. "Next time, you're going to have to tell me why you wanted white American cheese rather than the usual 'colored' yellow American cheese."

Now and only now did I realize how dark this man's skin was. I wasn't surprised he was black—it seemed most of the staff behind the counter in the mess hall were black—but now I realized how he was more mahogany, shiny with sweat, and big, really big. Then he smiled. "If you want lettuce, tomatoes, or onions, you get them at the salad bar. If you want fried onions, I got them here."

"Next time, I'll try the fried onions and the yellow cheese. This one is supposed to be an Uncle Chuck burger." I smiled back. A genuine smile and a genuine promise. I hope he knew I meant it.

The four of us sat at a table in a corner. The room wasn't crowded, as the rush had passed. Just a few tables occupied with soldiers lingering over coffee, smokes, and conversation. We stood out, being the only ones in uniform; the rest were in civvies, reducing the apprehension we still had of having a meal in a mess hall. Anyone with a stripe at Fort Dix was feared, because they could abruptly end our brief rest period, especially in the mess hall. That is, once you passed the entry test and got into the mess hall.

Every Army Basic Training mess hall has an entry test: the monkey bars. Yes, monkey bars, just like those in every school yard. To earn the right to eat, every trainee had to traverse the monkey bars. Problem was, my school yard didn't have monkey bars, so I had no experience. They looked easy. They were easy. Kids played on them, but there was a method, and I didn't know the method. Two pegs, one for each foot, about twelve inches off the ground. One foot on each peg, then launch your body, grabbing onto the horizontal bar, and hand climb across to the other side, maybe fifteen rungs.

I watched several of my fellow trainees before it was my turn. I mounted the pegs, jumped up, grabbed the first rung with both hands. No problem. I let go with one hand and reached for the next bar, crashing to the sand in the pit below. Must have slipped. I got the chance to watch the rest of the platoon attempt the bars, because the reward for my failure was to go to the end of the line and try again. Second attempt was no better. Pissed off, I tried six times. I wasn't the only failure. Three other trainees were having difficulty, along with me. After six attempts, the four of us were called worthless pieces of shit and told to double-time around the parking lot, a quarter-mile trot, before they allowed us to go into the mess hall.

After three days, failing each time before each meal, I was beyond frustrated. My fingers ached, and I had blisters. I was a former All-American athlete, and I was being beaten by playground monkey bars. A little skinny kid from Texas told me I was doing the monkey bars wrong and took the time to show me. I was a Harvard graduate, taking monkey-bar lessons from a seventeen-year-old high school drop-out who wanted to go to Vietnam to kill VC. He probably couldn't find Vietnam on a map, but he had a PhD in monkey bars.

It was evening chow, the end of a long hot day. We had no place to rush off to, just chow, and back to the barracks. He waited with me until the rest of the platoon earned their right to enter the mess hall. Two of the other three pieces of shit were still having trouble, so they were at the end of the line after failing on their first attempt.

"Ya see here, ya gotta think like a monkey. You don't see no monkey goin' one hand on the bar, then two hands on the bar, and then move one hand to the next bar, and then two hands on the bar. Shit man, even I can't do that, and I weigh a lot less than you, and I'm a lot younger," drawled Doctor Monkey Bars. "Ya just make your hand look like a hook and swing. One hand on this bar, other hand on the next bar, swinging your hips, ass, and shoulders, like yer dancing."

"Thought I was thinking like a monkey. Now I've got to dance?"

"Do you want to learn to do this, or do you want to eat all your meals in two minutes?"

"Please show me."

Easy as could be, this skinny little kid jumped up on the bars, hooked his hands and swung from bar to bar, ef-fort-less-ly.

Now, it was my turn. I jumped up, hands like a hook, and swung out, grasped the first bar with one hand, swung my hips, ass, and shoulder, and grabbed the next bar with the other hand. I was moving! Swung my hips and reached for the third bar. Got it! A new world record for me. Three in a row before I fell to the pit. I was elated. I was the best of the worthless pieces of shit.

"Hey, Drill Sergeant. I got it. Did you see? Three bars."

The Drill Sergeant stood rigid, arms crossed, feet spread at shoulder width. With his Smokey the Bear hat and sunglasses hiding most of his face, I couldn't see any emotion he might be showing. My tutor, my new best friend, stood at the opposite end of the bars where he had landed.

"Back of the line, trainee. You only got twelve more to go," the Sergeant growled. But I did see a nod of his head to the kid from Texas as he sent him into the mess hall.

I'll show him this next time. Five bars—another world record!

Instead of praise, the Drill Sergeant said, "All right, ladies. You keep falling off those bars, you gonna hurt something, and then I gotta whole lot a paperwork to fill out. Once around the parking lot at the double time. Now move it."

It took me three more meals and three more fast trips around the parking lot, but I did it. When I jumped off that final bar, I did a perfect Olympic landing, my hands up in the air, feet together, and I turned to the Drill Sergeant looking for acknowledgement of my new world record.

"About fuckin' time, college boy," was all he could manage. I guess he was as choked up and excited as I was. I did a little Arnold Palmer-like hitch of my fatigue pants and swaggered into the mess hall.

"No fuckin' monkey bars. That's what I really like about this place," I declared to no one in particular.

"No pussy is what I don't like about this place," replied Les to everyone in particular. "I like that place up on the North Post. Lots of pussy up there."

"No 'gansett is what I don't like about this place," added Kennedy.

"Let's go exploring and see if we can find us the enlisted club, and get us both what we need," said Les as he picked up his tray and headed toward the door, Kennedy right behind him.

I followed, reluctant to leave the first taste of sanctuary and sanity I'd had in almost twelve weeks—reluctant to head back across the street to our bedroom on the porch, but eager to be able to wander off and do anything I wanted to do by myself, for the first time in twelve weeks.

CHAPTER 7

"On our own again—sort of"

As we walked in silence back to the porch bedroom, I looked at the mess hall and assured myself there were no monkey bars. Not that they were a problem anymore, but like the skinhead haircut and walking in the sand on the side of the road, I didn't need a reminder of those times.

On the porch, I found it difficult to get settled. The place had no chairs, no beds or lights. With the sun setting, not enough light to read, if I had something to read. There was no television or radio. We fumbled around straightening our gear, but with no locker, it meant repacking everything in the duffle bags.

A strange thought ran through my mind: "Now that I have free time, I don't know what to do with it. For twelve weeks someone scheduled every minute of my life. The US Army brain-washing machine has made it so I can't think for myself anymore."

Les, back after learning there was no beer or women on South Post, stared at me. "Buddy, you okay?"

"Why?"

"You be talking to somebody about not having a brain anymore."

"Just a little tired from the long day. That's all. Too early to sleep. Just restless, I guess. Maybe a smoke and a walk will help."

"Now I'm sure you're not well. We been humping to and from everywhere for the past twelve weeks, and we finally have time to relax and do nothing, and you gotta go for a walk 'cuz you're tired and restless."

"Guess so. Maybe I'll find that brain you said I don't have anymore."

The screen door to the porch slammed as I walked out. It had one of those big springs to make it shut quickly. I hadn't noticed before, but now it seemed to emphasize my departure.

I started across the street toward the mess hall. The street had lots of foot traffic going in both directions. With the sun almost down, and the heat of the day dissipating, the night was pleasant. I walked, lighting an L&M as I went. The sound of hurried footsteps behind me caused me to pause and turn around.

"Want some company?" asked Kennedy. "Not much to do back on the porch."

"Sure."

I offered him a smoke. He paused after lighting it.

He began, "I quit smoking the month before I started Basic Training. At age twenty-five, I knew I had to get back in shape for the training and keeping up with the teenagers who were being drafted. Really had good intentions, but the only thing I accomplished was to give up smoking for the month. My exercise program started out that first day of not smoking, intending to go for a run. The run turned into a walk which I justified as just as good, considering I had been smoking for ten years. I also justified walking by recalling the physics formula for work, which didn't have a time factor in it. Work is mass through a distance. If I walked, I was still taking my body, the mass, through the same distance. Walking just took longer to do, but the work was the same. Should have pushed a little more, and wouldn't have been in such poor shape in Basic."

Turning to him as we walked, I said, "I didn't smoke until I got in Basic. I was an athlete, even played a little minor-league hockey before I got drafted. Once the physical training started, I didn't find anything difficult, except the monkey bars. So, when they gave us the mandatory break every hour with the 'light 'em if you got 'em,' I felt left out, and lit up."

Kennedy continued, "The physical part of Basic took a toll on me. All the running and standing got me. A lot of guys my age paid for their sedentary lifestyles with stress fractures of the feet. Mine happened in the fourth week of training, while on bivouac at the rifle range. My pup-tent mate and another guy got up thirty minutes early each morning, went to the cook tent, filled their helmet liners with hot water. They'd bring the hot water back so I could soak my feet enough to get them in my boots. Then, I'd lace those boots up tight for support, and suffer through the day."

"Why go through all that trouble when you could go on sick call?

"Why? To avoid a recycle back to the beginning of Basic. If you got a medical profile during the first four weeks, they recycled you."

"Oh."

"That first Monday of the fifth week of training found me at the head of the line for sick call. I got lucky and the doc who examined me took pity on me. After he found the stress fractures on the X-rays, he told me to lie down on a gurney while he checked something out. Nice old guy. Let me sleep for three hours, and then handed me a thirty-day Profile, the military equivalent of a doctor's excuse, limited my physical activity to 'no running, jumping, or prolonged standing for thirty days.' I celebrated by stopping at the hospital PX and buying a pack of smokes. That first cigarette after five weeks of no smoking made me so dizzy, I walked even worse on those stress fractures. I sat down. I didn't want anyone grabbing me and taking me back inside. The unknown of being admitted to sick bay twice in the same day seemed likely to carry the sentence of recycle."

There was a little vegetation on our left as we walked down the sidewalk. I could hear traffic rushing by at high speed up the slight hill, but hidden from view by the trees and bushes. On my right, behind the mess hall, there was a huge parking lot, more of an empty field being used as a parking lot.

The sidewalk ended at a well-lit tunnel. The tunnel being both the source of all the soldiers coming toward us, and the destination of the soldiers walking in the same direction we were. I thought there must be a PX or commissary on the other side. I field-stripped my L&M and put the filter in my pocket, and we headed into the tunnel, curious to see what was on the other side. As we moved through the tunnel, traffic moving directly overhead added sound and vibration.

Emerging on the other side, even though it was now past twilight, the area was lit up like daytime. Bright lights everywhere, including a helicopter landing pad to our right, and off to the left, beyond a well-manicured mall, a large parking lot. Bright lights on a huge building in front. It looked familiar and then I realized…

"It's the Pentagon," I mumbled to myself as I stopped dead in my tracks. Kennedy stopped, too. Foot traffic behind us moved around us. Foot traffic coming at us had to move around them. We caused a pedestrian traffic jam, fifty feet from the Pentagon.

Kennedy turned back toward the tunnel, "We probably shouldn't be here without some kind of pass."

"Shit," I mumbled and turned with him. Back into the tunnel I followed him. Fast, but not fast enough to pass anyone in front of us, not fast enough to draw attention. He didn't stop until we were back at the mess hall.

"We've been to the mess hall before and no one questioned our presence, so it's a place we're authorized to be," he said through pauses as he gasped for air.

I stopped alongside of him and lit up again. Leaning against the building, I took a couple of deep drags on the L&M and waited for him to speak.

"Jesus K. Christ, we could have really blown it," he mumbled, mostly to himself, but loud enough for me to engage him if I wanted. He continued, "Here we are with a chance to ride out the war at a desk in the Pentagon, and we go walking right up to it in the middle of the night. We're lucky the Military Police didn't catch us."

"Shit," I said, "we were lucky we weren't shot."

Les approached me with a fountain drink in his hand, looking concerned. "When you left on your walk, you was talkin' to yourself. Now, you're talking with him about being shot, and you doin' it in the rain, boy. You best come home now, both of you. Let's go."

I tossed the cigarette behind a bush, filter and all, and followed him across the street. He held the door open for us. As I walked in, he said, "That's better. You ain't talking to yourself anymore, you ain't talkin' about bein' shot, and you ain't standin' in the rain no more. I think I cured you."

"Cured who of what?" Dexter W. asked.

"West. I don't know, maybe just a temporary case of shit-for-brains."

Eyes were on me. Not knowing what they expected from me, I shook the water from my cap and brushed the few drops off my shoes. Then Kennedy broke in.

"We were at the Pentagon," he said. "It's right over there," pointing beyond the corner of the porch claimed by Les. "You have to go through a tunnel under the highway. It's less than two hundred yards from here."

"No shit," got echoed two times, then a third time as someone from the inside barracks walked through our bedroom into the night.

"Yes shit," I said.

"So what?"

Kennedy was flustered. "So what? I was over there without orders. They could have shot me."

"No fuckin' way. My girlfriend's sister came up here for a protest march two months ago, and they didn't shoot her," said Les.

"But I was less than fifty feet from the building."

"She sat on the front steps of the Pentagon, took a piss behind a bush, and they didn't shoot her. I think you're wrapped a little tight tonight. Take a shower, hit the rack, and get some sleep. You'll feel better in the morning," he suggested.

Good advice for all of us, all doable except for the part about getting some sleep. Screens aren't good protection from a heavy rain when it comes in sideways. Within an hour, we and all our gear squeezed into the one corner of the porch not getting wet. As I nodded off, I thought of the guys in Vietnam that night, sleeping in the rain and the mud.

But sleep didn't come easily. When it wasn't thundering and lightning, the heavy rain was beating on the roof of the porch. When the rain let up, there was a rush of traffic through the screen door— soldiers taking advantage of the break in the storm to get "home." Sleep snuck up on me as we lay waiting for the thunder and lightning to begin again, not realizing the last strike was the last strike.

CHAPTER 8

"Make Way"

The CQ had been right, there was no need for an alarm or a wake-up call. I fell asleep for a short while, but woke up to the screen door slamming, and slamming, and slamming, as the soldiers in the barracks began their day. Unable to sleep with the noise and traffic, I staggered into the latrine at sunrise, taking the obligatory shit, shave, and shower before returning to the porch to dress in the uniform of the day, fatigues. Les was still asleep when I got back, so I gave him a poke. In fact, I gave him lots of pokes before he even grunted.

"Les, rise and shine. We gotta get to breakfast, catch the bus, and be up at North Post by eight o'clock."

"Mornin', fellas. What a great night's sleep. It was just like bein' at Gramma's place up in the hills with all my cousins. All of us camping out on the porch, falling asleep listenin' to the rain fallin' on the roof," he said as he lay back with his hands clasped behind his head.

"What about the door banging all night, and people walking through?" asked Kennedy.

"Yeh, it was just like that. Shit, that door bangin' was nothin' like the noise my cousin Sarah Mae made when she was bangin' the guys who would visit her. Yes siree Bob, just like sleepin' on the porch at Gramma's house." Les rolled over and gave me a little wink.

Kennedy had stopped getting dressed and looked at Les, wondering, I'm sure, about cousin Sarah Mae.

We followed the soldiers taking the shortcut through our bedroom/ their porch to the mess hall. The early morning temperature was still comfortable.

In the mess hall, sunshine made the room brighter, emphasizing the contrast between the green and black tiles on the floor, and the sand-colored walls. The ceiling fans still turned slowly. The flowers once again welcomed us as we walked along the half-wall separating the entrance from the dining area. As I passed one of the vases, I could smell their fragrance even in the strong presence of coffee and bacon. Fresh flowers every day? In an Army mess hall?

This morning, a line forced me to a halt, looking at the backs of twenty soldiers ahead of me. The Army and Navy guys dressed in their khakis, and the Air Force guys wore their equivalents in blue. We were the only ones wearing fatigues. With more people than last night, there was more noise with louder conversations, chairs scraping, and the rattle of dishes. The line moved surprisingly fast. I was fifth in line from the soldier collecting money and looking at meal passes when I heard "Make way!" from behind me.

"Make way!" is the military term used to tell people someone of a higher rank approached, and because they're important, you had to get out of their way. At Fort Dix, everyone outranked us, so we heard "Make way!" all the time. I backed up against the wall in time to see a Green Beret Captain pass by and make his way to the front of the line. I glanced to my left at the line behind me and saw everyone leaning and looking forward with a smile, and giving each other curious little nods. Why the nods and smiles? Was I missing something?

When the Captain presented his orders at the desk, the soldier looked at them carefully, and then in a loud, yet respectful voice said, "Sir, is there a problem?"

"No, soldier. I'm just going to get my chow," said the Green Beret.

Coolly, and still with a lot of respect for rank in his voice, the soldier at the table said, "Sir, the end of the line is back there."

"Yes, soldier, I know. I'm going to the front of the line."

"Sir, no disrespect intended, I'm sure, but—"

"No disrespect taken, soldier," said the Green Beret.

"Sorry, sir, I'm sure you intended no disrespect, but this is the South Post ENLISTED Man's Mess Hall." The soldier at the desk emphasized "enlisted." "While we welcome visitors of all branches of the military, and all ranks, we ask you to RESPECT the fact you are a visitor and must abide by the rules of the mess hall, which are posted at the entrance. Sir, those rules state all those who wish to use this facility must have orders, a meal card, or cash available for entry. Those not in uniform must be prepared to show a valid military ID card."

He paused, looked at the Captain, glanced down the line of stretched necks connected to grinning faces, and continued with his practiced recitation. "Sir, those rules further state senior non-commissioned officers will not invoke their privilege of moving to the front of the line, but take their position at the end of the line. Sir, it further states officers will abide by all the above rules, as guests of this enlisted man's mess. Sir, in this case, it means you must go to the end of the line.

Alternatively, there's a snack bar back near the officer's quarters, where you can get bad coffee and a stale donut."

"Soldier, I'm a Captain, and for the eight years I've been in the service of our country, I've never had to stand behind a Private in any mess hall. Let me speak to the NCOIC or the OIC in charge of this mess hall responsible for this breach of military etiquette."

"Sir, the NCO in charge is Master Sergeant McQueen, and he will direct you to the notice from General Hadley, Post Commander, at the entrance to the mess hall."

At this, a Master Sergeant approached the table.

"Captain, please uncover in my mess hall." Uncover is the Army term for "take your hat off." Turning to the soldier at the table, he continued, "Why is this line not moving, soldier? People have to eat and get to work."

"Master Sergeant, the Captain didn't get a chance to read the directive from General Hadley as he came in, and I was explaining it to him."

"Your job is to check for meal cards or collect money, not be an information booth." The Master Sergeant turned to the Captain and said, "Sir, if there's something you don't understand about the policy of this mess hall, then may I suggest you go to the entrance, read the directive, and decide whether you want to eat in my mess hall."

Dead quiet surrounded me now. I couldn't hear the necks straining or the smiles stretching. I could hear everything on the other side of the half-wall, but on this side, silence.

The Captain was not shaken. "Sergeant Major, I have an appointment at the Pentagon—"

Before he could finish, the Sergeant Major interrupted, "Sir, most if not all the soldiers behind you are permanent staff at the Pentagon, some of whom handle the assignments of Captains. You may even feel the need to ask one of them for a favor today, and sir, they will tell you what I'm telling you now: 'Sir, all the rules, all the time.'"

The Captain put his orders back in his pocket and turned to face the gauntlet of soldiers he had passed on his way to the front of the line. As he walked to the end of the line, virtually every soldier greeted him with a smile and either "Sir" or "Captain," including two Majors standing in line behind Les, one of whom turned and shouted, "Make way!" as the Captain moved to the end of the line.

Les learned from one of those Majors this was a long-cherished ritual, enjoyed by all who used the South Post Mess Hall. When an officer or senior NCO, unfamiliar with the local rules, entered the mess and started down to the front of the line, rather than telling him politely the local rules, someone would shout out "Make way!" and watch the fun begin. Most, after having been initiated, recognized the good nature of the ritual, and halfway to the back of the line were smiling and accepting the ribbing.

At the front of the line, I showed my orders. Last night, the mess hall had the distinct aroma of dinner. This morning, smells of coffee and bacon dominated. Steam tables offered a wide assortment of breakfast selections: oatmeal, and next to the oatmeal, a white pasty dish I learned to be grits, a staple of the southern breakfast. Then there were waffles, pancakes, hot cinnamon buns, biscuits, gravy, syrup, bacon, link sausage, patty sausage, little steaks, and thick slices of ham. Next to the ham rested scrambled eggs, fluffy and golden yellow. I had taken pancakes, biscuits, syrup, and bacon, and found myself standing in front of the eggs wondering if I really wanted eggs, when a cook asked me if I'd prefer an omelet.

"Sorry," I said, "Did you say omelet? I don't see any omelets here."

"I make omelets to order," he said. The smile on his face acknowledged my station in life here as "newbie." "Made to order, two eggs or three, and you pick what you want in it."

"What have you got?" I asked, realizing I was playing the same game I had played last night with the cook, and looked forward to being a happy loser again.

He waved his spatula across the tops of little stainless-steel tubs that contained about a dozen omelet fixings: cheese, onions, peppers, ham, pieces of bacon, chopped-up tomato, mushrooms, and some stuff I didn't recognize.

No scrambled eggs for me. "Let's go with a two-egg omelet, bacon, mushroom, onions, and cheese, please." Two trips to the South Post Mess Hall, and I knew already I'd be saying a lot of "please." When he handed me my perfectly cooked omelet a few minutes later, I also realized I would be saying "thank you" a lot.

Last night's salad bar was now a cold breakfast bar, with several kinds of breads and rolls, fruit, cold cereal, and juice. At the end of the cold breakfast bar was the coffee station. I grabbed a mug, some sugar and cream, and headed to the nearest table for four.

There was more noise and conversation this morning compared to last night, and the morning diners ate with a purpose. Conversation was more animated than relaxed, and with few exceptions, after they ate and the last of the coffee drained from their mugs, everyone got up and moved on with the rest of their day. A few singular diners remained enjoying a second cup of coffee, a smoke, and reading the newspaper.

The others joined me at the table. Kennedy had taken advantage of the omelet station, but Les and Dexter W. each had a plate of biscuits and sausage patties covered in gravy, and a side bowl of grits.

"Man, I ain't had no grits in almost three months," Les said as he put six pats of butter on the grits. He stirred the butter in as it melted, and then laced it with salt and pepper. "Only thing better would be some red-eye gravy for these biscuits. The boy behind the counter told me they only got red-eye gravy on Sunday morning, 'cuz there's less folks here. Takes a lot a time to make. Guess where ole Les is havin' breakfast on Sunday? I might put some of that there cheese in my grits on Sunday, just to make it a real special day."

We didn't respond to him, unless you count the grunts as we ate. When we finished, no one was in a hurry to leave. We still had thirty minutes until the bus took us to whatever adventure awaited us on North Post. Kennedy and I enjoyed a cigarette with a second cup of coffee.

"What do you think we're going to have to do up at North Post today?" asked Kennedy.

"Looks like we're goin' have to do all the soldiering work these guys that's all dressed so pretty ain't gonna do." Les licked the big soup spoon he used to eat his biscuits, sausage, and gravy.

"Hey, Les," said Kennedy. "You started telling us earlier about your cousin who used to bang all the guys on the porch…"

"I didn't start to tell ya nothin'. I started and finished telling ya what

I was gonna tell ya. There ain't no more to tell."

"Yeh, but was she pretty? Did you ever do her?"

"Boy, Sarah Mae is kin. What I said is all I gotta say about it." Les' tone was on the pissed off side of neutral.

An awkward silence followed. Kennedy looked embarrassed. He punched out his cigarette in the tin ashtray, got up, saying he wasn't used to eating so much, and should probably go take another shit. Dexter W. left with him. That left me with Les, both of us with a cup of coffee, a freshly lit cigarette, and no urgency in our bowels.

"Seems like you're real close with your family," I said. "All those cousins and trips to the mountains."

Les looked me straight in the eye and said, "I don't have a lot of cousins. I don't have a Gramma who lives in the mountains, and if I did have a cousin Sarah Mae, I sure wouldn't sit back and let her do a gang bang." The southern accent was still there, as was the magical twinkle in his blue eyes when he spoke; but the hillbilly twang and the "ain'ts" and "gonnas" were gone, replaced by an educated southern gentleman.

"What have you done with Les?" I asked.

He smiled. "Most northerners I've met seem to think if someone's from the South, and they talk with a definite accent and talk slowly, then we're all hillbillies and marry our cousins. I played the poor ole country boy at Fort Dix, and it kept me out of all the college boy crap you got from the cadre." Les took a drag of his smoke and a mouthful of coffee. "Most of those guys at Dix were high school drop-outs, so it wasn't much of a challenge. So, I figured I'd have some fun 'wit yous college boys,' and feed your stereotype image. Even being educated, you're still pretty gullible.

"I come from an upper-middle-class family in Atlanta, went to Georgia State, started working in a bank while waiting to hear from a law school, when Sam got me for two years. Don't blow my story, okay? I'm having fun."

I was grinning from ear to ear. He had me fooled, too. Clever man, taking on a new persona with a new group of people in a new stage of his life, and having fun with it. This new phase of our lives, and the uncertainty, scared most of us to death; but here he was, enjoying himself.

"I had a feeling this morning, when you gave me that little smile when you talked about your cousin, there was something going on. Then, when you did the sausage, biscuits, gravy, and grits, I was back to believing you. I mean—grits! That's going a bit far."

"Oh, don't go messin' wit' my grits, now," said Les, back in hillbilly mode. "Grits is the mother's milk of the southern boys, and there ain't a finer breakfast anywhere than sausage, biscuits, and red-eye gravy."

"You're good, Les. Your secret is safe with me."

We finished our coffee and cigarettes, took our empty trays to the washing area, and set out to face our adventure at North Post.

CHAPTER 9

"Back South to North Post"

Kennedy and Dexter W. were already waiting at the bus stop. Three other soldiers, also dressed in fatigues, waited, all looking like prepubescent boys on the first day of middle school. None were looking at each other; some were kicking pebbles into the street, or picking imaginary pieces of lint from their uniforms. When Les and I joined them, all we got was a nod.

Soon, I fell into the same funk. Yesterday morning we were at Fort Dix in our Class A uniforms, waiting for a bus to take us to the Pentagon and the start of a great adventure. Twenty-four hours later, we're back in fatigues, waiting for a bus to take us into the unknown. Fatigues reminded us of all that time in the sand on the side of the road, marching, sweating, and being shouted at by morons. Fatigues reminded us we were still in the U.S. Army.

The bus arrived, and we piled on, dead quiet for the entire ride. No questions about the north vs. south dilemma. We got off the bus where we had started our Fort Myer adventure yesterday, in front of the Consolidated Barracks. Still with lots of time, we strolled to the back of the building, entering on the side near the big parking lot. The CQ who had made our "room reservations" the night before wasn't there, but an equally nasty Corporal had taken his place.

"What the fuck do you sorry-ass mother fuckers want?" he demanded. We looked at each other. We didn't know what we wanted, and we no longer knew who was in charge, after yesterday's CQ had stripped me of my responsibility. When I looked at Les, he shrugged as if to say, "Not me little buddy, you got the orders."

I stepped forward and said, "I am, Corporal," and handed him our orders.

"You must be the college boys that called in for room service last night. I heard about you. Shit, we all heard about you. Which one of you is Kennedy?"

Jim stepped forward and said, "Here, Corporal."

The Corporal looked at him, gave a shrug, and told him to get back in line.

Turning back to me, he snarled when he spoke.

"You am what, college boy? I axed you what you want, and you say, 'I am.'" How the fuck did you get outta college, givin' wrong answers like that?"

"I meant to say I'm in charge, Corporal," I said.

"No, you ain't, college boy, I'm in charge now. And for the time being, I'm gonna make this here big, blond white boy with the smile in charge when I'm not around," he said as he approached Les. "What's your name, college boy?"

"My name is Leslie, Corporal. Please to make yer acquaintance," said Les in his best corn-pone, hillbilly twang, stepping forward with his hand out to shake.

"How the fuck did you get outta college? You talk like Gomer Pyle. I gotta look into the G.I. Bill and go to college. If you sorry-ass mother fuckers can get outta college, anybody can."

Turning from Les, he continued, "Okay Gomer, get those orders from the dickhead that use to be in charge, and give them to me." Les reached over and took the orders.

The Corporal moved in front of Kennedy. "They say I gotta be nice to you, 'cuz you a Kennedy from Massachusetts," he said with a wave of his arm. He screwed up his eyes and looked Kennedy up and down, saying, "I don't think so. If you was one of them, you wouldn't be here with me. And even if you is, I don't give a shit. I live in Alabama, and dey ain't done jack shit for me, and I'm getting out in three mother-fuckin' weeks."

The Corporal moved away to face the rest of us. "Today, and for as long as it takes you, you are going to police up this here ground floor of the barracks. It's called the ground floor 'cuz it's under the ground. That little brown line on the walls ain't a decoration. It's the high-water mark of the flood, week before last. Rain run down the ramp from the parking lot into this here ground floor. The Army Corpse of Engineers, another bunch of worthless college boys, didn't make our parking lot too good. Because they didn't want any rain water running down the hill on the other side of the fence into the Cemetery, they just tilted the parking lot toward this here barracks. Thing is, these college boys didn't put the drains in the right place, and the water comes here faster than it can get to the drains. So, it's you college boys who gots to clean up after dem college boys. You gonna do the walls first, so it looks and smells pretty where I gotta work. Then, when that's done, you gonna start cleaning all the gear back in the supply rooms, and dat is a lotta shit. We gots about a hundred rifles that are all gonna have to be broke down and cleaned and oiled and put back tagether. Now, don't that sound like fun?"

Wanting to moan, groan, and beat this mean, miserable little mother fucker to a pulp, we instead said, "Yes, Corporal."

"Wrong answer again, college boys. You think you gonna have fun, but you is wrong. I'm gonna have fun watchin' you. Now, get those hats off in my house, and start washing my walls," he ordered.

We scrambled down the hall. The second door we opened had cleaning supplies. We grabbed mops and brushes and filled a couple of buckets with water and detergent.

"You college boys is sumptin'. You grabbed floor mops, when I said you was gonna wash my walls," he shouted at us. Les already had a wet mop wrung out and was starting at the wall.

"Mops work jes fine on walls, Corporal. Back home, we use them all the time for stuff like this. Sometimes, even when a goat or a pig backs into the wall, and we gotta clean the goat shit or pig shit off the wall. Momma says it's better than usin' a brush, 'cuz the brush can take off the wallpaper; but the mop is so smooth, all it does is make it peel a little. We just put some new glue on it, and it's as good as ever."

"You have goats and pigs in your house?" asked the Corporal.

"Yessir, Momma don't let us bring the cows in anymore. Cows have a lotta gas, 'cuz a all the grass they eat, 'specially in the spring, when it's new grass. You gotta be careful with the oil lamps and the cook stove when they's around. Goats and pigs are okay, tho, 'cuz they don't eat as much grass, so the farts ain't so big as to start a fire," Les added.

I had a tough time not laughing.

It seemed to work, though, because the Corporal was scratching his head and mumbling. "Make sure you don't take no paint off a my walls with those brushes. Use dem mops."

We made pretty good progress, especially by Army standards, finishing the main office of the CQ, and were moving into the hallway as the clock showed ten a.m. The Corporal hadn't said a word since Les had described his home life, but now he shouted for him.

"Hey you."

"Yes, Corporal," responded Les as he ran to the Corporal., stood at attention with the mop at right shoulder arms.

"Around here, the Army says you get a break every hour. It's ten o'clock, and you ain't had one. Tell your guys to take a break, and be back here in twenty minutes."

"Where can we go?" asked Les.

"Anywhere you want, as long as you be back here in twenty minutes."

"Can we go to the mess hall and get a cup of coffee?" asked Les.

"Can you get there and back in twenty minutes?"

"Don't know. That's why I asked you."

"Get the fuck out of my sight, or you ain't gonna be in charge no more."

Les took charge. "Everybody out. Break time for twenty minutes."

When we got outside, I burst out laughing and started to say something to Les about the act he was putting on, only to get a stern look from him that reminded me the others were as much fooled by him as the Corporal. Kennedy agreed to get us a coffee while we tried to get through three cigarettes in twenty minutes. All but Les, who asked him to get a sweet tea or a Dr. Pepper or a cherry Coke, instead of coffee. If they didn't have sweet tea or Dr. Pepper or cherry coke, he'd have a coffee.

Standing alongside Les, I complimented him on his performance.

"Yeh, it seems to be working. Ol' Corporal hasn't bothered us much since he started thinking about cow farts in momma's cooking stove."

"No sweet tea, whatever the fuck that is, and no Dr. Pepper, so you got coffee with cream and sugar," Jim said, as he handed Les his coffee. I took my coffee from him just in time to start my second L&M. I finished my third cigarette in twenty minutes as I walked down the ramp into the basement of the Consolidated Barracks.

The rest of the morning went by without incident. The Corporal seemed content we were getting the crud off the walls without messing up his paint job. Les continued the hillbilly act, and the Corporal stayed at his desk with his coloring book and crayons. We all had a dull look on our faces, that same dull look I had seen on the faces of all the soldiers at Fort Dix who were doing meaningless work. The dullness in the eyes that pondered, How long do I have to wash these pots and pans, or how long do I have to peel potatoes, or how many times do I have to mop the floor, or how many times do I have to run around the parking lot? Once again, we faced the unknown, and it got us down on ourselves, especially bad after the uplift in spirits we had experienced yesterday.

As noon approached, we were anticipating a big "Hey you," telling us we could take an hour for lunch. Our progress was slowing, and we focused our efforts on the wall by the supply room. When the "Hey you" came, we weren't going to waste any time. We'd be close enough to drop our mops and buckets and rush out the door. As we each chose a painted cinder block on the wall to give our special attention to, we noticed someone had come into the CQ.

The sunlight from outside cast the newcomer as a dark outline against the outside backlight. As he came further into the room, we could see he was another soldier, dressed in a tropical-blend, green Class A uniform, not the typical khaki everyone else was wearing. His service cap was on, and the polished black brim reflected the overhead light, making the dark shadow appear to emit a point of light from the middle of his head.

"Take your hat off in my house, soldier," said CQ without looking up from his comic book.

Taking off his service cap, the stranger said, "That's 'take off your hat in my house, Command Sergeant Major.'"

Corporal CQ dropped his comic book and stood up at attention, knocking his can of Coke over. He righted the can and made a feeble attempt at cleaning up the mess, but the stranger said, "Let me help you with that." The stranger reached across the desk, grabbed the dropped comic book, and used it to brush the spilled Coke into the garbage can. Then he took the other comic books and used them to sop up the remaining soda, throwing the wet mess into the garbage can. "I guess comic books are good for something, aren't they, Corporal?"

"Yes, Command Sergeant Major. What can I do for the Command Sergeant Major?" squeaked the Corporal.

The stranger stepped forward and we could see him more clearly. An older man, perhaps fifty, he was of medium height and build. His head was almost shaved, and what hair remained was grey. Pink cheeks and pale complexion were a stark contrast to soldiers who spent their time outside. He had a chest full of ribbons, fitting someone of his rank. A Command Sergeant Major was the second highest-ranking non-commissioned officer, with only one man outranking him: The Command Sergeant Major of the Army.

The stranger opened a manila envelope and took out two sheets of paper. He looked at the papers and then looked at the Corporal.

"Yesterday, three soldiers left Fort Dix with orders to report to me today. They have not arrived yet. Do you know where they might be?"

"Command Sergeant Major, da three soldiers are working for me taday, TDY, doin' some clean up after da storm."

The stranger looked at the paperwork in his hands and said, "I don't see any authorization on these orders for my three soldiers to be assigned temporary duty with you." He passed the orders to the Corporal saying, "Do you see anything on these orders assigning them temporary duty to you?"

Without looking at the orders, the Corporal said, "No, Command Sergeant Major. There's nothing on dem orders about temporary duty, but we been doin' dat wit the guys comin' in to work at da Pentagon."

"Well, Corporal, we ain't gonna do dat wit my soldiers no more. Where are they?"

"They be right der in da hallway, workin' on the walls, Command Sergeant Major."

"Get them," commanded the Command Sergeant Major.

"Hey Gomer, get you and you boys out here, on the double."

We put our mops, buckets, and brushes in the supply room, trotted to the CQ office, and stood at attention facing the stranger. He slowly walked over to us, looked us all over, paying particular attention to the name tags sewn on our blouses. He stopped in front of Les.

"Is your name Gomer, soldier?"

"No, sir."

"Not sir," he said, pointing to the chevrons on his sleeve. "I work for a living. Address me as Command Sergeant Major."

"Yes s—, I mean yes, Command Sergeant Major."

"Do you read comic books, Private?" he asked Les.

"My mother is New Orleans French Creole, so the recreational reading I was allowed to do was the French classics, in the original French," responded Les, with a decidedly southern accent, but with none of the hillbilly twang.

The stranger said something to Les in French, and Les responded to him in French, and they both chuckled.

The stranger turned his attention to the Corporal. "Corporal, these soldiers are through with this TDY assignment. I will be sending a directive to your commanding officer that will state very clearly, in the future, soldiers assigned to me will report directly to me."

"Yes, Command Sergeant Major," stammered the Corporal.

The Command Sergeant Major turned to us and said, "I am Command Sergeant Major Stoner, the NCOIC, which means Non-Commissioned Officer in Charge of the office you were to report to today. That office, EPCMR-GS, is responsible for all 7, 8, 9, and 0 series MOSs for grades E-6 and below. We make the worldwide assignments for about four hundred thousand soldiers. The Corporal here probably has a clerk or supply MOS, so he is my responsibility. I can send him anywhere in the world—tomorrow! Even someplace where they don't have comic books." He continued talking but moved to stand in front of the Corporal. "If any of you are selected to work for me in my office, you will have the same control over the Corporal."

The room had been silent except for the Command Sergeant Major, but as he finished, a mop fell to the floor, probably hiding the sound of the Corporal's asshole slamming shut because he puckered so hard.

I liked this guy, this Command Sergeant Major. He was taking our side, not because he liked us or because he didn't like the Corporal, but because it was the right thing to do. Someone had been bending the rules, using soldiers in transit to do the menial work that was supposed to be the work of the general duty soldiers. The Corporal would get the message back to the rest of the rule-benders that it wasn't going to happen anymore, even without the directive to the CO, HQ Company, U.S. Army.

Still looking at the Corporal, Stoner said, "I assume that because you've had time to read comic books and make work assignments for my men, you have also provided them with adequate billets until we make their permanent assignments." He looked at Les for confirmation.

"Command Sergeant Major. They got us all set up on the porch down at South Post. It isn't good, but it's a whole lot better than some of our boys in Vietnam have to deal with," offered Les.

The Command Sergeant Major turned to us and said, "We're going to take the bus back down to South Post, where you're going to have a nice lunch in the mess hall. After that, you soldiers will gather your gear from the porch"—turning toward the Corporal—"and then you will return here to the Corporal by 1500 hours, at which time he will provide you with your room assignments in this building."

"But, there ain't no rooms—" stammered the Corporal, but he was cut off before he could finish.

"Corporal, I was in this Army when everyone wore brown shoes— over thirty-five years. I know the drill. You may not have any 'rooms,' but you have room in the rooms for my soldiers. Take a look at your room roster, and see where you have two of your buddies in a six-man room. Put my soldiers in there. Tell your buddies to make them welcome, because when they report to me on Monday morning, I'm going to ask them how their roommates are. If I hear there are mean roommates, I might just have to see if they're ready for a rotation to Vietnam, where they can be mean to the VC. Got it?"

"Yes, Sergeant Major."

"That's Command Sergeant Major, Corporal," corrected Les.

"Yes, Command Sergeant Major," repeated the Corporal.

Stoner continued, "Gentlemen. After you get your room assignments in this building, settle in, make friends with your new roommates, get familiar with Fort Myer, and relax. Enjoy your weekend, visit our nation's Capital just across the river, visit Arlington National Cemetery, or just do as you see fit; but report to my office at 0800 on Monday."

He then nodded to us, turned, and walked out of the office. His shadow could be seen standing in the hallway.

Corporal comic book broke the silence. "You heard what the man said. Go git that bus, git some chow, git your gear, and be back here at 1500. Now git outta here, I got work to do," said the Corporal, every word said looking at the shadow on the floor in the hallway.

Les took charge. "You heard the man. Let's get moving."

Putting the cleaning gear away, we saw the shadow in the hall disappear. We followed the shadow into the hot noon sun. The noon heat of Washington in the summer would never feel as good as it did that day.

As we walked to the bus stop, we were different again. The funk and the dullness that had invaded our bodies this morning was gone, replaced by the spirit and hope we had yesterday. There was a bounce to our step, and we were engaged in conversation. We looked like the people around us. Had we become soldiers, or were we merely back in the human race? The singular event that transposed us was the appearance of the Command Sergeant Major, taking us from cleaning up someone else's mess, getting us a room at the inn, and giving us the weekend off. We no longer had to walk in the sand on the side of the road.

Lunch was casual, and except for the fatigues, we fit in with the rest of the soldiers at South Post mess. We were even a little more casual and relaxed than the others. While they had a limited time for lunch, we had a couple of hours. It wasn't like we had a lot of packing to do. All we had to do was walk across the street and pick up our duffle bags, already packed.

CHAPTER 10

"Television and Toilets"

At a quarter to three, we presented ourselves to the North Post CQ. With this guy being such a prick, we didn't want to take the chance we'd miss the deadline, giving him a reason to give our rooms away. The remnants of lunch cluttered his desk; a half-eaten burger, fries, and a Coke, all his major food groups.

"Make sure you dickheads tell the Command Sergeant Major I gots you all rooms here in this building. It weren't easy. Make sure you tell him." When he got no answer, he looked up at us. The high sun of noon was gone, and with it, the intense glare from the front door. The muted afternoon sun softened the concrete block walls and metal desk.

Les enjoyed being singled out by the Command Sergeant Major, and continued in his role as our leader. "If the rooms are okay, I'll be sure to tell him when I see him on Monday."

"Good," came the quick response from Corporal CQ.

And with his best Gomer smile, Les continued, "And if any of us has a problem with our room, I'll let him know that, too."

Corporal CQ knew he was beat. The presence of the Command Sergeant Major remained in the room from earlier in the day. He became ordinarily human. "I'm gonna issue you two sheets, a pillow, a pillow case, and I'm gonna issue you a blanket. The rooms here is air-conditioned, and it get chilly at night. You decide if you want to take this extra bedding now wit your other gear and go to your rooms, or do you wanna come back for it after you settle in?"

We looked at each other, but before any of us could answer, Les said, "Let's save us a trip. Give us all our bedding now."

The elevator came, and we piled in, duffle bags, hangers, and bedding. At the fourth floor, the doors opened and Corporal CQ said, "All out."

Corporal CQ led us down the hallway, stopping about halfway. Producing two keys, he gave me one and the other to Les. "Make sure it works." We tried our keys, and both worked. "Don't lose it. Cost you five bucks to get a replacement."

Corporal CQ left, with Kennedy and Dexter W. following. Les called down the hallway as they departed, telling them to meet us for dinner at the mess hall at five o'clock.

The room was the size of the average college dorm room, with two sets of bunk beds. Both the upper bunks had been made up, leaving Les and I the choice of a lower bunk or a lower bunk. Two desks, with metal chairs for each, huddled together. The chairs didn't look comfortable, which probably explained why the desks didn't seem used. Instead of closets, there were four wall lockers in the room. Unlike the trusting openness of the South Post Barracks, two lockers had combination locks. I examined the other two and found them empty, except for two wire hangers in one and an empty Dr. Pepper bottle in the other. I took the wire hangers and told Les, "The other one seems to be reserved for a Southern boy."

Les chuckled and responded in corn pone, "Yes sirree. Sure do look like this here is the best locker in the room. Already come decorated by Dr. Pepper hisself."

Before we unpacked our duffle bags, we decided to explore, partly out of curiosity and partly because I needed to take a leak. The bathroom was the next door down the hallway.

The bathroom was different from any others we had seen in the Army but fit in with the rest of the college dormitory style of the building. Tile walls and floors, it had eight urinals against one wall, and six enclosed toilets further down. Across from the urinals, eight sinks had mirrors above them and fluorescent lights everywhere. Further down and around the corner, we found another room containing ten individual shower stalls.

Individual toilets and individual shower stalls! What a treat compared to the transient barracks where we processed in at Fort Dix.

The transient barracks appeared to have built during the Civil War, and the bathroom added after the invention of indoor plumbing. It was a dim room with urinals on one wall, and on the opposite wall similarly vintage sinks, where we shaved and brushed our teeth. The urinals and sinks were old with cracked porcelain and rusty fixtures. Above the sinks hung mostly de-silvered mirrors. There was a room for toilets: a room about twenty feet by twenty feet, with toilets around three walls. Around the top of these three walls were cute little six-panel windows that swung out at the bottom from high hinges. It was the kind of window you might expect to see in a Norman Rockwell print, if Norman Rockwell ever painted an Army latrine.

The floors were wide wooden planks worn smooth and shiny after decades of use. The walls were bare, no sheetrock or boards. Bare studs separated us from the outside board wall, where patterned shards of light outlined the old timber. The light from these cracks augmented the light from the low-wattage bare bulbs hanging from the ceiling.

There were no stalls, just toilets. And in front of each of the toilets was open space, a lot of open space. The room was roughly 400 square feet. Each toilet took up about six square feet. With nine toilets on each of the three walls, that's only 162 square feet, leaving more than half the room empty. Sitting on the toilet on one of the wing walls, you looked at nine guys sitting on the toilet trying to take a shit, while you were looking at them watching you trying to take a shit. If you were shy and weren't into group dumps, it could be a problem. Not much of a problem for the first couple of days for me, or a lot of other guys for that matter. The Army diet, the exercise program, and lack of sleep had a constipating effect. The Army knows about this and figured it out. New recruits are too busy the first few days to have time to take a shit. Then about the third day, or maybe it's the turd day, the Army issues you a shit. After all, they've issued you everything else you need—sheets, blankets, clothes, boots, and after three days, it's time to put a megaton charge of atomic poop shooter in the breakfast tomato juice.

Problem is, they don't tell us we're going to be issued a shit. They let us sleep fifteen minutes later than usual, and scheduled twenty minutes of barracks time after a somewhat leisurely breakfast, even with time for a casual walk back to the barracks. Actually, the first part of the walk was leisurely; the second part became a sprint because the atomic poop shooter raced into our colons and started looking for a way out.

This particular morning, I chose to eat quickly, relishing the thought of twenty minutes of barrack time, free time. That choice put me at the head of the crowd when my colon told my brain to tell my legs to move faster and get me to the latrine. I wasn't the first to get there, but let's say I was in the first seating. I dropped my trousers and had a projectile dump that splashed my ass. My first G.I. shit was the only thing the Army had given me all week that seemed to fit and felt good. Around me, others were experiencing the same relief.

As a civilian, such defecation gratification was something to be enjoyed in private, undertaken at leisure, many times accompanied by a book, maybe an interim flush, and then a few more minutes on the throne to make sure the enemy within was indeed without. Not so as a soldier in the transient barracks. Why? Remember all that space in the four-hundred-square-feet room? The space wasn't wasted. It was the waiting room. Because while the first seating was on the throne, the rest of the guys who wanted to be seated were standing there. Well, not just standing there, they were dancing the "green apple two-step." They were screaming at you, rooting you on, cheering those who flushed, cheering those who reached for toilet paper, cheering those who stood.

It had all the elements of the last two minutes of the championship basketball game in a small gymnasium I had watched as a teenager. Only this time, I wasn't a spectator, I was on the team! Not just on the team, but playing in the big game. A cheer goes up as I reach for the toilet paper, wadding up a big handful of the thin, almost transparent Army-issue ass-wipe tool. I reach behind with one swipe—cheer—a second wipe, and the crowd roars. As I reach for another wad of paper, the crowd groans, but immediately comes alive again as I wipe only once. I smile at the cheering crowd as I stand and pull up my G.I. issued boxers. The crowd is rushing toward me. I turn and flush just as the skinny little kid with the Texas drawl gets to me. Expecting praise, I wait to be lifted on his shoulders. He says, "About time, mother fucker."

"Yes," I say to Les, "Toilet stalls," reflecting on how far I had come since the nadir of spectator sport defecation. It also made me humble and thankful, as I thought of the thousands of soldiers who would take a shit in a hole in Vietnam today.

We explored a little before heading back to our room to unpack. A laundry room sat next door to the bathroom with four washers and four dryers, coin-operated, with a vending machine dispensing little boxes of detergent.

Further down the hallway an open lounge area welcomed us with comfortable looking chairs, sofas, end tables, a coffee table sort of thing, a card table, and a television.

"Man, this is just like my college dorm," said Les as he walked to the television and turned it on. We waited a few seconds for it to warm up. The few seconds turned to a few minutes, until a thin horizontal line appeared across the screen. Les sat down, put his feet up on the sort-of coffee table, and started watching the line.

"Les. What the fuck? It's a line," I said.

"It's TV, and I ain't seen no TV in almost three months."

I sat down next to him to watch the line.

Two guys came into the area from the other side and saw us watching the line. "Television's broken," said one. "Has been for a month."

"Thanks, fellas," said Les in the corn pone again, "but I ain't seen no TV in almost three months, and this is just fine for me. How many channels do we get?" He got up to change the station.

"We don't get four stations," said the voice, "ABC, NBC, CBS, and the public access station, all out of Washington."

"What numbers are those stations, boys?" asked Les as he stood at the television.

"We don't get Channel 4, 7, 12, and 13," said the voice.

"Thanks," said Les, as he changed the channel, stepped back, looked at it and said, "Ah huh," changed the channel again, said, "Ah huh," and changed the channel the final time. He stared at his final selection, the same straight line we first had, scratched his head, and changed the channel back. He smiled at the straight line, turned up the volume so he could hear the hiss, returned to his seat on the couch, and put his feet back up on the sort-of coffee table.

Another guy came into the room and asked, "What are you watching?"

Les responded, "Channel 7."

"What's on?"

"I don't know. Just got here."

He sat down on the couch with Les, put his feet up on the sort-of coffee table, and glued his eyes to the straight line.

I told Les I was going back to the room to unpack, and reminded him we had to meet at the mess hall later.

"Okay, buddy. I'll be along just as soon as this is over."

I wanted to ask how he knew when the straight line would be over, but I was afraid his answer might just convince me it wouldn't be long, and I'd sit down and join him.

After I unpacked my duffle bag and stowed my gear in the locker, I took a leisurely walk to the bathroom, cleaned up, and put on my khaki uniform. I waited in the empty room for a few minutes. Being alone was good, but spooky. It was the first time in months I had been in a room by myself. I needed people. I went back to the common area to find Les and his new buddy still staring at the line on the television, not saying a word to each other, and joined them.

CHAPTER 11

"On the outside, looking in"

Dexter W. was waiting for us when Les and I arrived at the Consolidated Mess Hall for dinner. While we wore our Class A khaki uniforms, he wore civilian clothes, except for his Army-issued black shoes.

"What the fuck! We had to turn in our civvie clothes at Fort Dix," said Les.

"Yeah, we did, but we had that weekend pass after four weeks of Basic. When I went home, I brought this stuff back. They didn't say I couldn't, so I did."

"But where did you leave it?" I asked. "They inspected our lockers every week."

"When we got ready for an inspection, I'd put my civvies in the dryer in the cadre's laundry. Figured they'd think it was one of their guy's stuff. After the inspection, I'd grab a mop, slop their floor a little, get my stuff, and put it back in my locker."

Les and I recited the "Son of a bitch" chorus together, like we had been practicing it for years.

"Friday must be date night," said Les, noticing there were fewer people in the mess hall than the previous night. We learned later that weekends at Fort Myer were like some small colleges in New Jersey. Lots of people live on campus, but head home on the weekends for Momma's cooking, Momma's laundry, and to make sure Jody wasn't making any headway with the girlfriend. Here, most people managed to leave the office around noon, jump on a train at Union Station, or get in the car and be home for supper.

Dinner proved to be a pleasant time for us to reflect on the progress made during this eventful week, think about relaxing over the weekend, get settled in, and get ready for our big day on Monday, when we started work at the Pentagon.

"First thing on my list for after dinner is to go to the PX and see if I can get some civvies," I said.

I got a nod of agreement from Les, but Dexter W. said, "As soon as we finish eating, I've got to get the bus to Union Station to catch the train home. Gonna see my wife, see my folks, get more clothes, and bring back my car."

"Don't you need a weekend pass for that?" asked Les.

"The Command Sergeant Major told us to enjoy the weekend. Didn't say anything about staying on the post. In fact, he suggested we take in the sights of Washington and Arlington Cemetery. If we can go to Washington, we're leaving the post without a pass. My way of enjoying the weekend is to go home and get my car."

I don't know what Les was thinking, but I was thinking about how lucky I was to be here and not in Vietnam, and I wasn't going to do anything to jeopardize that. I would seriously consider staying on post all weekend if there was any doubt.

"Dexter, if you're not supposed to be doing this, and if you get caught, don't tell them you told us, okay?"

"Told you what?"

"You know, that you went home," I responded, failing to get the message.

"I didn't tell you I was going home. Where the fuck did you get that idea?"

"Didn't you just say—"

"Forget it," Les said. "Leave it alone."

Dexter W. left the table first. Les and I lingered over coffee and small talk. There was a sense this was the beginning of the end of us as a group. We had shared a thrilling and emotionally uplifting experience for the last thirty-six hours. Our fate, and maybe our lives, had changed when they plucked us out of Fort Dix, avoiding a guaranteed ticket to Vietnam. None of us knew each other before meeting at the bus station yesterday morning, but we bonded at once because of our good fortune. Now, we were starting to disassemble. Corporal CQ split us up in our room assignments. Dexter W. left the first chance he had. Kennedy disappeared without saying anything to anyone. The next time we'd be all together was Monday morning, when we reported to the Command Sergeant Major who would decide our fates. Would we all be assigned to the same place, or would they assign us to different places? Monday we'd find out.

"I hope Dexter W. knows what he's doing," I said. "I hope he doesn't get caught, if he doesn't know what he's doing."

"Doin' what?" asked Les, with the same smile that accompanied the corn-pone accent.

"Yeh, doin' what?" I said. "Right now, we make a pact. If Dexter W. gets caught, and we get asked about it, we don't know nothin'. Agreed?"

"Agreed."

We had taken sides, us against the truth. Two of us would lie and throw Dexter W. under the bus to protect our asses. I felt bad, because I liked him.

The agreement to lie, if necessary, threw a wet blanket on our conversation, so we gathered our stuff and took the dirty dishes to the cleaning area. Outside, I headed for the PX to see what I could do about getting some civvies. Les was of a similar mindset.

We took a right out of the Consolidated Mess Hall, and then another right onto Sheridan Road, the main street of Fort Myer. All the streets on Fort Myer had been named for dead Generals. After we passed out of the shadows of the modern Consolidated Barracks and Mess Hall, we took a trip back in time.

On our right stood four red-brick buildings that looked like they had been built at the turn of the century. These buildings housed various agencies assisting the soldiers assigned to Fort Myer. These buildings gave way to a massive parade ground the size of a dozen football fields. Halfway down the parade ground, near the sidewalk, stood the flagpole of Fort Myer, officially designated as Building #1. It had to have a building number so it could have maintenance performed on it. On every Army post, they always designated the official flagpole as Building #1.

This ground had been in use for almost one hundred years. All the land Fort Myer sat on had been confiscated from the Lee family and turned into a military post following the Civil War. The parade ground had seen soldiers from every war since then. The first and second airplane flights on a military installation took place on this parade field, both piloted by Orville Wright. Sadly, the third flight, on September 17, 1908, resulted in the first airplane crash and the first fatality. Orville was at the controls again, and his passenger, First Lieutenant Thomas Selfridge, died. The government named Selfridge Airforce Base in Michigan after him.

Across the street on our left was the home of the Third Infantry Division, or Third Herd, that we had glimpsed from the bus on our trips to and from the South Post.

Four buildings stood side by side, all identical in size and appearance. Two-story, red-brick structures built out to the street, with only room for a sidewalk in front. Each building included a wooden porch on the first and second floor running the length of the building. Individual rooms opened directly onto the porch. On the porches, soldiers looked busy and serious, as they should. I heard no idle chatter, and no music from the open doors, even though I stood only fifty feet away. They may have participated in two or three funerals today, and were preparing for tomorrow's funerals. Several might even be getting ready for their turn at guarding the Tomb of the Unknown Soldier.

Les and I continued our walk in silence.

At the end of the Third Infantry Barracks, there was a short, one-block street off to the left, ending at the main entrance to Fort Myer. We could see traffic on the highway outside of the fort, so we crossed the street and walked to the entrance.

There it was, the outside world—the civilian world we had been a part of three months ago. The street we saw was Arlington Boulevard, also known as Route 50. Across the street was an apartment complex of two-story buildings. Cars were parked in front, somewhat in disarray, so they weren't part of the military. We could hear kids playing on the sidewalks, ridings bikes, shouting at each other. What a sensation! After three months, we were back in contact with the "world." There were real people living real lives, wearing real clothes, driving real cars, and living right across the street.

The MP posted at the entrance waved cars in, saluting those carrying officers, and answering questions for those that stopped.

"Buses don't stop here," he called to us. "You have to walk down to the next corner, or cross the street."

"What?" I asked, realizing he was talking to us.

"Come here," he said, using his white gloved hand to signal the same message.

We walked over to him.

"I said, the buses don't stop here, you have to walk down to the next intersection, or you have to walk across the street to catch a bus. You guys just got here, right?"

"We're not waiting for a bus," I said. "How did you know we're new here?"

"First, you got that fresh-out-of-basic-training look in your eyes, not the look of someone who's been to the 'Nam. Those guys show up with the thousand-yard stare. Second, you're walking around with time on your hands, so you're not in the Third Herd. And third, you gotta be the only two guys on this post wearing your uniforms, other than the Third Herd—so you just got here, probably today."

"You're wearing a uniform," challenged Les.

"Correction, soldier. You got to be the only two soldiers on the post wearing a uniform who aren't working."

"Yeh, we got here yesterday from Fort Dix, and we got the weekend free before we start work at the Pentagon on Monday," I said, trying to gain a bit of dignity after being targeted as a newbie.

"If you plan on going into the District," he said as he waved another car in from the highway, "I'd recommend you ditch the uniforms and get civilian clothes. Even so, with those haircuts, you'll still stick out among those long-haired hippie freaks, but—"

"We don't have passes to leave the post," I said.

"Shit, man, you are newbies. You still have that 'scared shitless of rules' attitude."

"Yeh, how'd you know?" asked Les.

"You don't need a weekend pass to go to the District, or for that matter, just about anywhere. It'd be a good idea to get a pass if you wanted to take a train or a plane. That way, you show your pass, and you might be able to fly military standby, or get a discount on a bus or train ticket."

"How's that? Are there two sets of rules—one for Fort Dix, and one for the rest of the Army?" I asked.

"There's two sets of rules, okay, but it's one set for the rest of the Army, and another set for the guys here at HQ Company, U.S. Army,

Fort Myer. See, most of the guys assigned to HQCUSA work at the Pentagon, because they're needed to keep the 'Five-Sided Paper Factory' operating at full capacity. If you leave it up to the civilians, the building would sink from all the paper coming in and none of it going out. The enlisted men are all college grads, with a few in graduate school or law school. So, they figure these bright guys will find ways to fuck things up if the Army sticks them with the chicken-shit rules and regulations. They just tell them this is a bad-paying job you've got to do for two years. You're kind of a super-civilian, so they treat you like a civilian. You work eight to five, five days a week, and the rest of the time is yours."

"No shit!" said Les, with my "no shit" sounding like an echo.

"There's other soldiers working at the Pentagon, 'lifers,' who are the cream of the crop. They don't like to see the college boys getting away with all the stuff they do, but they didn't get where they are by bucking the system. And, these lifers take care of their men. Just a different set of rules here."

I remembered the Command Sergeant Major who came to our rescue earlier today.

"We already met the lifer in charge of us. Seemed like you said, taking care of his men. So, the trick here is to think like a civilian?" I asked.

"Sort of. More like, 'think like a civilian, but act like a soldier.' You don't understand that now, but it'll all become clear to you. Then, you'll have me to thank, and I'll have friends in the Pentagon."

"You got that right, friend," said Les, with his biggest and best southern corn-pone smile. "And we got us a friend in the MPs if we need him, right?"

"Friend, if you work in the Pentagon and get into any trouble where you need a friend in the MPs, you'll be in so much trouble your friend in the MPs isn't going to be able to help you. But, if you want to think that, you go right ahead and call me when you're in trouble."

We got our new friend's name and contact information, gave him our names, and promised to contact him as soon as we had a telephone number.

Before we left the guard house to continue our trip to the PX, we walked across the street and stood in the civilian world. Sure, we had been in the civilian world traveling from Fort Dix to Fort Myer. This was different. Intellectually, per our new friend, we understood we weren't breaking any rules; but emotionally, we felt like we were AWOL, on the outside in the civilian world, looking across the street at the Army beyond the gates and the fence.

CHAPTER 12

"Cars, stars and Hershey bars"

Les and I returned from our short immersion in the civilian world refreshed, and continued our walk on Sheridan Road. All U.S. military installations, be they Army, Navy, Marine, or Air Force, are small towns, providing all the necessities, and some nicer things, to make life livable on post. Behind us, at the south end of Sheridan Road, sat the commissary, which is the Army name for supermarket. Next to the commissary was the gas station, which, besides pumping gas, had six service bays where soldiers could do their own repairs. The bays included lifts, grease hoses, and drains for oil. For bigger jobs, you could borrow heavy equipment, like engine hoists, and jacks. Across the street from the commissary complex was Radar Clinic, which housed the medical and dental facilities, including a basic emergency room.

We approached the "town center" of Fort Myer. Here was a post office—small, non-military, and looking very much like they had transported it from some small town. A movie theater stood next to the post office, also non-military and transported from the same small town, probably on the same truck. The marquee listed a movie I had seen before I left home, more than fourteen weeks ago.

As we came to the head of Sheridan, it turned sharply to the right, becoming Jackson Avenue. Across the street on the left stood the majestic officers' club. This was neither non-military nor small town. The building oozed military and authority! After all, it was the "O club" for most of the Generals who ran the Army. A swimming pool sat behind the O club. We crossed the street to take a closer look at how they took care of the brass. While kids played the pool, they seemed to be more reserved in their play, with less noise than the municipal pool where I grew up. For kids, less noise meant less fun.

Les and I had been silent since our brief trip into the world beyond the gate, communicating with nods and shrugs. Les broke the silence. He seemed to read my mind. "Nice pool, but them there kids don't look like they're having as much fun as me and my cousins used to have skinny dippin' in Gramma's crick."

"That's 'cuz they don't have your cousin Daisy May skinny dippin' with them," I said, trying to imitate the corn-pone accent.

"Gomer, you got two things wrong there. Number one, you got the worst southern accent I ever did hear. And two, my cousin's name wasn't Daisy May."

Around the corner from the O club we found two other buildings. One was our destination, the PX. The other, the NCO club, with a small bowling alley attached. Still feeling the freedom of the outside world, I offered to buy Les a beer at the NCO club.

"I'll accept your offer as an apology for the poor imitation of my native tongue, and your failure to remember my cousin's name."

The cool dark embraced us as we entered, each a dramatic change from the outside. The coolness was immediately refreshing, while the blindness that happens when you leave the bright sunshine and walk into a dark bar took a few minutes for adjustment. Even without sight, there was no mistake it was a bar. The same stale cigarette smell, and the smell of more than one beer dropped on the floor that crept into the cracks between the tiles and wood. A smell that never left, but got fortified regularly. Standing in the cool dark, we waited until our eyes adjusted to the dim light, then walked to the bar.

"What'll you have?" I asked Les.

As the bartender approached, Les answered and ordered at the same time "Two cold beers, pardner," he said to the bartender.

The bartender was like any other bartender in America, except for a few things: He had a G.I. haircut, and he had rules. "Sorry pardner, but I can't serve you."

I reached for my ID card to prove I was old enough to drink in any state having a drinking age of twenty-four or older. As I put the card on the bar, Les caught on and reached for his.

"Ain't about age, it's about rank. This is the NCO club—that's E-4 and above. You boys just got those E-2 butterflies on your uniforms, so I can't serve you." He said it politely, and with the right amount of matter-of-fact that meant we shouldn't even think about starting an argument about it.

"Damn," said Les. "I was looking forward to having my first beer in a real Army bar. If we can't go to the O Club, where's a fella like me supposed to get a beer?"

"Could go to the 123 club."

"What's that?" I asked.

"Well, you got the O club for the officers, and you got the NCO club for the E-4 and above, and you got the 123 club for the E-1, E-2, and E-3s," he answered.

"Well, there you go—our own 'Privates' club," quipped Les, amused with his joke. "Where might that be?"

"The closest one is down in Fort Belvoir, about twenty-five miles south of here."

"Shit man, that don't seem fair, us having to go all the way down there for a beer, when you got a really nice club here."

"Ain't about fair, soldier, it's about rules. The reason we don't have a 123 club is we don't have many privates. Most of the soldiers stationed here have been in for a while. Most even have had a tour in the 'Nam and come back with some stripes. Even you boys who come down to work in the Pentagon, fresh out of Basic or AIT, get promoted quick. So, if I was you and I wanted a beer at this here club, I'd either get promoted real quick, or I'd get out of those uniforms. Any soldier who comes in here in civvies, we assume he knows the rules. Ain't nobody ballsy enough to come in here that don't belong. We don't bother askin'," he said, with just the faintest hint of a smile.

We thanked him, apologized for breaking the rules, and left. Outside, we stopped to finish the smokes we had started in the bar.

"You know," I said to Les, "I'm getting a feeling our jobs at the Pentagon are pretty special. The Command Sergeant Major kicked Corporal CQ's ass today for us. We seemed to get special treatment when we were talking to the MP at the gate. Now, the bartender tells us how to get into the NCO Club for a beer. What do you think?"

"Little buddy, I think you are just not used to how friendly folks down here in the South are. I didn't know southern hospitality extended so far north." Looking at him, I couldn't tell whether he was corn-poning me, or stating a fact as he believed it. The look on his face could go with either.

We finished our cigarettes on the way to the PX. I had never been in a PX, not even at Fort Dix. After I was lucky enough to get my life-altering physical profile that eliminated any prolonged marching, walking, or standing for thirty days, I became the company's permanent orderly. After that, my training consisted of cleaning the barracks and latrines most days, while the able trainees learned how to become the "Ultimate Weapon." The only breaks I got from that duty were when they woke me up at 0300 to go on KP duty. Some mornings, they didn't have to wake me, because they had assigned me to man the supply room overnight. Sleep was at a premium, so using any free time to visit the PX never entered my mind.

There's no such thing as a typical PX. Each one takes on the flavor and needs of the community it serves. For instance, not much call for long underwear and parkas in Fort Polk, Louisiana. Basically, it's the local department store. It has a little of everything, and always at a better price than the discount stores, like Two Guys and Korvettes. There's always a jewelry section with rings and watches. It's not Tiffany's, but some of the folks shopping at a PX think Tiffany's is a breakfast place in New York City where Audrey Hepburn eats. There's always a big electronics section, with lots of 35mm cameras, radios, hi-fi sets, portable televisions, and 8-track players. But none of these were on our shopping list; we were looking for civilian clothes.

Les found the clothes department. Actually, he found the ladies' clothing department and the men's section was right beside it. Les had his back to the men's department and was talking to a nice-looking young brunette. She pointed over his shoulder toward the men's department. As I approached them, he turned around, following the length of her arm with his nose, his face close enough to her hand he could lick it. He opened his mouth. Shit, he was going to lick her hand!

Corn pone to highest degree, he said, "Now don't that just beat all. I should a seen that clear as day, but I had this woman in my eye, and I couldn't see nothing else 'cuz of the dazzle of her beauty."

She raised her other hand, the left, and brought it around to his face. "See this dazzle, Private?"—waving a couple of rings, including a big shiny diamond on the ring finger. "This must be what blinded you. It means 'married' most places, but here it means married to an officer. So back off."

"Yes, ma'am. Sorry, ma'am. Thank you kindly for the directions."

She turned her back to us, continuing her shopping at a rack of blouses. We walked twenty feet to the men's department.

"Boy, Gomer," I said, hoping he would recognize I was giving him a new name whenever he went corn pone. "Looks like you struck out. She wasn't interested, and she was married, and she was married to an officer."

"Can't hit the ball unless you swing the bat," he replied. "Mickey Mantle was a star batting .300. If I only batted .300, I'd give up, or worse yet, I'd have to get married, just so I could get laid regular. I saw she was married, it's the first thing I look for. Maybe she was lonely and the only thing she had to do was go shopping. Anyway, that was just practice. I been out of practice for the whole time I was at Fort Dix. It'll come back. I ain't worried."

"I can't wait to watch the master score with that 'dazzle in my eye' shit."

We each went through the clothes, Banlon shirts and khakis for me. A Banlon golf shirt would suit me fine now, and double as a golf shirt later. I chose khakis for pretty much the same reason. I caught up with Les, who had a blue madras-print, short-sleeve shirt with a button-down collar, and a pair of white slacks.

"Is that the best you can do?" he asked, looking at my selections. "You've been in the Army for three months, wearing green most of the time. When the Army lets you dress up, they let you wear khaki. So, what's the first thing you do when you can buy civilian clothes? You buy a green shirt and khaki pants. Are you some kind of secret lifer?"

The selections seemed a poor choice, considering what he said. "But I want a shirt I can wear playing golf later," I said as I stood in front of the shirt selection, undecided. Les came over and pulled a dark blue Banlon golf shirt and handed it to me, without saying a word. He moved over to the pants rack and pulled a pair of white-on-white bell bottoms.

"Here," he said, giving me the pants. "Yeh, it's blue and white, just like mine, but it ain't like we're going on any kind of double date or something."

Suddenly, I was ten years old again, shopping at Grant's Department Store with my mother. Only this time I was twenty-four, and Mom was a corn-pone talking member of the Georgia pussy posse.

Still wanting a bathing suit for the pool this weekend, I choose a pair of black boxers. Les nodded moderate approval. Not a bad choice; they only had black!

Even though we finished our shopping, we continued exploring the store. Fort Myer would be our home for a while, and knowing what was available would be helpful. We wandered on the outskirts of the kids' clothes and the baby section. Off to the side, they had a catalog desk, where you could order lots of other things, including big items, like refrigerators and sofas.

What we found that was a surprise was a section of the store selling Army stuff. Now, I don't know why that should have been a surprise, but it was. After all, this was an Army base, and stuff wore out. I assumed if something wore out, like a pair of shoes or boots, there was some place you could go, and the Army would give you another pair. Same thing with uniforms. Maybe there was, but here they had everything we had in our duffle bags, and then some.

"Shit, if I had known they had all this stuff here, I would have just taken my shaving kit and underwear, and left the rest of the crap in Fort Dix. I could've gotten all new here, and not have had to lug that duffle bag. Prices are good, too," Les said.

I guess that attitude goes with coming from money. Les was willing to drop a couple of hundred dollars to replace everything in his duffle bag, and his duffle bag, for the convenience of not having to carry it.

"Must be nice to have that kind of money," I said.

"It ain't about the money, Gomer."

"Why am I Gomer, now?" I asked.

"When you say something that ain't smart, you're a Gomer, so I'll call you Gomer," was his answer.

"Why ain't that smart?" I asked.

"'Cuz, all the shit you've got in your duffle bag is shit. You got three pair of fatigues you wore about forty times each. You wore them eighteen hours a day, in the rain, when you crawled in the mud, when you crawled in the sand, and when you did KP. They are plumb wore out. Same with those boots you got. If you leave those boots in your locker for a week, they're gonna dry out, curl up the toes, and crack. So, you got all those green fatigues and those dirty boots you're gonna throw away."

"What about these khakis we're wearing? We can't wear the same pair every day."

"These are shit!" he said emphatically. So emphatically, the brunette he had "practiced on" in the ladies' department looked our way. "They make these from surplus WWII tents used in the African desert campaign. Feel how thick and heavy they are. Now feel these," he said, turning to khaki uniforms on the rack behind him.

I reached over and felt the uniforms on the rack. "Jesus K. Christ! These are fantastic! They feel fantastic," I said as the uniform I was wearing started to weigh me down and make me itch. "What is this?"

"It's a wash-and-wear Class A uniform, my friend. Drip dry. No more ironing! If you don't spill anything on it, just rinse the stink out in the sink when it needs it, shake it out, let it dry overnight, and you're ready to go in the morning. And I'm going to get me one, right now."

"Me too, Mr. Gomer," I added. We'd been at Fort Dix for three months, gotten three months' pay, and had no opportunity to spend it, until now. "Might just buy two."

"Good idea, Mr. Gomer. Let's get two, and throw this shit out tonight."

We searched the rack, and each found two uniform trousers and two uniform blouses. Blouse is the Army name for a shirt! I don't know if I was more excited about getting back to the barracks to put on this new uniform, or the civilian clothes.

We continued looking around the military section of the PX. In an alcove on the back wall, we found a display with all kinds of brass. Brass is the term used to describe the shiny gold-and-silver-colored stuff the Army makes you to put on your uniform. The Army issued us two pieces of brass for our collars back at Fort Dix. Both were a stamped bubble with the U.S. Army logo on one, and something else on the other. It was cheap and hard to clean. The brass comes varnished, so it won't tarnish, but they make you take the varnish off. With the varnish removed, it tarnishes, so we'd have to polish it every week with Brasso. Fancy two-piece brass hung on cards on the shelf, brass like the NCOs wore at Fort Dix.

"Let's get these to go with our fancy new uniforms too," said Les.

"Roger that."

They also sold various military patches, medals, and rank insignia. This is where we would come to get our new rank insignia when we got promoted to PFC. As we moved down the display wall, we came to the officers' section. There were officer uniforms, lightweight like we had just picked up. The green Class A uniforms weren't much different from ours, except for the black stripe running down the outside of each pants leg. Big difference in the hats! The brimmed hats for officers are distinctive in having gold braid on them—a single gold braid band for Lieutenants and Captains, and additional gold "scrambled eggs" on the brim for the higher-ranked officers.

Tucked in the corner, we found a military jewelry store, with gold and silver mementos to go with each rank. I guess when an officer gets promoted, the folks who work for him have to give him something to celebrate the occasion. But then again, based on the egos of the officers I had seen at Fort Dix, I could see them buying this crap to decorate their home offices or garages, or to give to Mom and Dad.

In the far corner of the jewelry section, we saw an incredible sight—something that could only be seen in this PX, home of more Generals than any other military base in the U.S., or maybe even the world.

Officers wear jewelry as rank insignias. Lieutenants get either a single silver bar or a single gold bar, Captains get two gold bars, Majors get a silver oak leaf, Lieutenant Colonels get a gold oak leaf, and full Colonels get an eagle. Jewelry sections in other PXs probably had everything we had just seen. But now is when we knew we were at Fort Myer, because tucked in this corner were the rank insignia for the Generals—one star, two stars, three stars, yes, even four stars. At the very top, on the far left, hung the circle of five stars for the General of the Army. There was not just one set of these, designating the very highest-ranking General in the Army, but six sets of these five-star circles.

"Shit, man. Look at these," I said as I lifted one set off the rack. "I don't think there's been a five-star General since Eisenhower."

"Wrong," corrected Les. "Marshall was the last. Must be why they got so many left here. Need someone to get promoted to move the inventory. They are awful pretty. I know a little blond lady in Savannah who would like one of these to hold her pony tail. Before you get any ideas about that lady, Gomer, she's my eight-year-old niece."

Les took the card with the circle of stars from my hand, looked around, and said, "I guess we're done with this here store for now. What say we head back to the barracks and try on our new gear?"

"Roger that," I replied.

Les got to the checkout counter and put his uniforms, collar brass, and the five-star circle on the counter. To comply with the sign at the counter, which advised all customers must show a valid military ID or dependent's ID, he was pulling his wallet when the clerk said, "Congratulations seem to be in order, General—especially for achieving such distinction at such a young age. May I see your orders, please?"

Les pulled his orders assigning him to the Pentagon, and presented them along with his military ID card.

"No, sir, I need to see the orders promoting you to General of the Army."

Les looked at her with concern. "I ain't no general. This here circle of stars is for my niece's pony tail."

"Sorry, I can't let that happen, Private. If I sell you these stars without orders, I lose my job. If you buy these stars, you could go to Leavenworth Army Prison for impersonating an officer."

Les went into corn pone. "Can't have that happen to either of us, can we? I guess I'll just hafta tell her about the pretty circle of stars I was gonna git her."

"Smart move, Private. Just so you know, you ain't the first. If I could sell those five-star circles without orders, I'd be running out every week."

Les paid his bill and moved to the front of the store to wait for me. I settled with the clerk without comment, and caught up with Les as he exited the building.

"I guess we still got a lot to learn about this here Army," he said. "You think they'd have some signs around the place. 'Don't come in this bar unless you are such-and-such a rank.' 'Don't buy this unless you got orders.' But we are learning a lot this here first day, ain't we, Gomer?"

"We certainly are, Ollie," I responded without thinking.

"What's with the names? First you call me Gomer, then Roger, and now Ollie?" asked Les.

"I don't know why I said that. Just flashed into my head. The line from is the Laurel and Hardy movies. When they'd get into a fix, Ollie would say something, and Laurel would reply, 'We certainly are, Ollie.'"

"I don't get it," said Les.

"Maybe you'll be able to see an old Laurel and Hardy movie on the TV in the day room this weekend, and then you'll get it."

"I forgot all about the TV. Man, what a day—new civilian clothes, new military clothes, and TV," he said, as we headed back down Sheridan to our new home.

Back in the barracks, we tried on our new clothes. With both of us in blue-and-white, the civvies seemed like another uniform. I chose to wear my bathing suit and a white undershirt to spend the rest of the evening reading old magazines in the TV room, while Les sat watching the horizontal line.

The weekend proved to be less than we had hoped for. We donned our civvies on Saturday and headed into Washington, only to realize our short hair made us stand out as much as if we were walking around in uniform. Even more so when we went into Georgetown.

Georgetown seemed to be the collection point for the East Coast hippies that weekend. Our short hair made us potential targets for the anti-war protestors. Mother nature must have recognized our sorry state and started a rain storm to beat all rain storms, sending us looking for a cab to take us back to Fort Myer. With our civvies wet, I reverted to the bathing suit and T-shirt, dreading another evening of stale magazines in the television room. Miraculously, someone had replaced the broken set with one that worked. Les was in heaven, and I watched, too.

The barracks was eerily quiet. For the past three months I was used to sharing everything with sixty other soldiers. The few who remained had weekend duty while others seemed to have part-time jobs as judged by the fast food uniforms they wore.

Sunday morning was a treat. Les slept in. Our two roommates were among those who left for the weekend. I strolled over to the Consolidated Mess for breakfast. I considered going down to the South Post for the special Sunday morning breakfast, but it was either raining again or still raining, and waiting for the bus at both ends of that trip made little sense.

Piles of newspapers sat near the tray pickup area. Some just thrown there, others reassembled after reading. I grabbed a reassembled one, got my breakfast, and settled down to have a civilized breakfast, reading my newspaper and smoking my cigarettes over coffee, for as long as I wanted. My first meal alone in three months!

Les arrived two newspapers, three cups of coffee, and three cigarettes later.

"Just woke up in a room by myself for the first time in three months," he declared.

"Just had my first meal by myself in three months," I responded. "This might work."

"Just might," he said, heading toward the serving line. He returned with biscuits, gravy, grits, and coffee. He grabbed a newspaper and started his breakfast.

"What are we going to do today?" I asked. "If the rain lets up, we can go down to the Kennedy grave."

"Rain ain't gonna let up, Gomer," he said, sopping up gravy with the remains of a biscuits. "Looks like our first day of nothing to do." He stopped mid-chew and looked toward the door. "Lookee, lookee. I see nookie coming to save my day."

I turned around and saw Les' day-savers. They weren't all day-savers, but I got his drift. There were six young women lining up to get their trays. Les and I watched them as they moved through the line. They pulled two tables together and sat.

"They aren't saying much to each other," I said.

"That's the trouble when you get a couple of lookers in with a lot of dogs: nobody talks. But seeing as they don't have nobody to talk to, I think the nice thing to do is help them out. Don't wait up for me, Gomer," said Les as he grabbed his tray and made his way to their table.

It wasn't noon yet, and Les was telling me not to wait up for him!

I watched long enough to see they didn't chase him away. After seating himself near the homely end of the table, he worked his magic and his seat toward the prettier girls as the others left. I really could learn something from this man.

I returned to our room in the barracks, straightened things out, and spent most of the afternoon watching television. Dexter W. came in around 5:00 p.m., back from his trip to Rhode Island. He insisted I go out to the parking lot to see his car.

"It's still raining," I said, "It can wait 'til later." I could tell he was disappointed, so I gave in and went out back. The rain had let up a bit, and we walked to the middle of the parking lot.

"There she is, the sweetest '68 Chevy Malibu on the planet," he said proudly.

It was a sweet-looking car. Black with lots of aftermarket chrome. Sitting there in the light rain, it looked like black ice. "Nice," is all I said.

"And quick," he said, rambling on about the 396 engine, Weber carburetors, the four-speed transmission on the floor, rear end, the tires, and the dual exhaust system. He pointed to the "ice cube trays" on the front hood with particular pride.

I like cars and have had a few nice ones, mostly foreign sports cars, but I'm not a gear head. He could have been talking Greek, for the understanding I had of the details he recited. I managed a "Nice," or the occasional "Sweet," all said at the right time and with the proper amount of respect, because he beamed each time.

The rain became heavier, forcing us back inside. On our way, we passed the ramp that had taken us to the basement supply room and CQ yesterday. CQ had been right about the Army Corps of Engineers' job. A lot of the rain that had fallen on the four-acre parking lot was now water rushing down the ramp into the supply room.

"Seems you should be able to put something across the top of that ramp to divert the water away," I said, translating a random thought to words.

"Yup, seems you could," was all he said.

Dexter W. and I went to dinner at the Consolidated Mess. The girls from breakfast came in during our meal. That is, most of them came in during our meal. Two of the better-looking women were absent, and there was no sign of Les. I grabbed a Hershey bar on the way out, heading back to the barracks. The anxiety was setting in about the next day, our first day at the Pentagon.

CHAPTER 13

"Hello, Pentagon"

Monday morning started at 0530, not because I set my alarm early, but because my body woke me, anticipating the big day—the start of my assignment at the Pentagon. The latrines were empty, so I had a leisurely shower and shave, and got back to an empty room by 0600. Les had not returned from his day with the two women from the mess hall. I put my new uniform on, buffed my shoes again, and headed over to the Consolidated Mess Hall for coffee to kill time.

The morning was glorious after rain the previous night, the sky a brilliant blue, with just a few wispy clouds, and the wet grass a vibrant green. With the temperature still low, it proved the start of a great day. The mess hall was empty, except for a few people in civilian clothes. We had agreed to meet at the bus stop at 0700, take the bus to South Post, and have breakfast there. Being closer to the Pentagon, a short walk from the mess hall assured us we wouldn't be late on our first day.

Kennedy and Dexter W. had beaten me there, greeting me at the big stainless-steel coffee urn.

"Guess you couldn't sleep either," said Dexter W. as he finished filling his mug.

"Yeah, big day. Didn't want to miss it."

"Where's Les?"

"Don't know. His bed hadn't been slept in."

"He's a big boy. He knows what's at stake."

"I hope so," I replied, filling my mug and heading for the cream and sugar.

We took a table near the door and settled in, no one speaking. After two smokes, I needed a refill on my coffee. Looking at my watch, I noted only twenty minutes had passed. This would be a long hour.

At 0630, the mess hall began to fill up. Most of the soldiers were dressed for work in summer khakis.

When I returned to the table, Dexter W. noticed my new uniform.

"Where the fuck did you get that uniform? Is it regulation?"

"Les and I picked these up at the PX over the weekend. Don't think they'd sell them to us if they weren't regulation." Then I told them about how the sales girl at the check-out counter had kept Les from buying the circle of stars.

"Hmm, must be legit. They look nice. Might have to get a set. Have to see how they make it through the day before I spend any money, though," Dexter W. said.

The mess hall continued to fill as the clock crawled towards 0700. At 0645, I decided to head for the bus stop.

"No sense taking a chance," I said as I got up. "Besides, all the smoke in here is bothering my eyes. Might be nice to get some fresh air while we can."

"It's your fucking smoke," declared Kennedy.

I sent him a look, the look of one smoker to another, and challenged, "What's your point?"

The three of us cleared the table and walked to the bus stop. I was nervous. On the outside, I appeared relaxed and casual, but inside was something else. This was a big deal.

When we arrived at the bus stop, I lit another cigarette from the stub of the one I just finished.

Dexter W. looked at me. "Nervous?" he asked.

"Yup. How about you?"

"Yup. But you can't see my asshole's puckered up. I can see you chain smoking."

At five minutes to seven, I put out my seventh cigarette since 0530, and looked toward the far end of Sheridan Drive, toward the commissary, for the bus. As we stared into the distance, we heard a yell behind us.

"Hey. You wasn't going to leave without me, was ya?"

Les came running up the sidewalk, almost dressed. He wore his new uniform, but had not yet tucked his shirt in, or put his belt on, and he was carrying his garrison cap. Besides getting our attention, his call attracted the attention of a Staff Sergeant on his way to breakfast. Although not close enough to make out the words, it was obvious Les was getting chewed out for not being fully dressed. Les stood straight as he tucked his shirt in, put his belt on, and put his garrison cap on, at first backward, but quickly corrected. As the bus approached, Les became animated, saluted, then extended his hand for a shake before running to us at the bus stop. The Sergeant ignored both his handshake and salute.

Out of breath, he got to us as the bus arrived.

"Almost missed the bus, Les," I said.

"Almost don't count. Besides, just like women, if I missed this one, another one or two will show up right behind it." This he said with a big smile and a wink for me. "If you get my drift, little buddy."

"You're right," I acknowledged. "We're early. There's still an hour before we have to report. Five-minute ride to South Post mess, twenty minutes to eat, and five minutes to walk to the Pentagon. Lots of time."

"See, and you was worried about me. Nothin' to worry about."

While Les fussed with tucking his shirt in as he got on the bus, I asked. "Les, don't you worry about catching something from these women you sleep with?"

"Nope."

"When I was in Basic, they gave us the VD talk before the first weekend pass."

"Yup, we got one too."

"But I didn't need it. When I was growing up, two guys from town joined the Air Force in a buddy program. After four weeks of basic training, they got their first pass, and got the VD movie. One of them, Benny Wurtzberger, was so shaken, he vowed not to leave the post all weekend. Benny's buddy convinced him to come along on a trip across the border to Nuevo Laredo, about two hours south of Lackland Air Force Base—a rite of passage at four weeks. Benny agreed, but said he wouldn't go into any bars. When they got to Nuevo Laredo, everyone bolted for the first bar they saw. Benny followed, but only after vowing not to drink. He didn't want to get drunk, and he didn't want to take the chance he'd get some disease from a dirty glass or bottle. Then he vowed not to drink unless they gave him an unopened beer bottle, so he could clean the top and open it himself. The next thing he remembered was waking up back at Lackland, in his own bunk, with a terrible taste in his mouth."

Les sat next to me on the bus and gave me his full attention for this story. "Too much tequila?"

"That's what he thought, until his bunkmates showed him the Polaroids, eight of them, only part of his face visible, but his name tag clear. Each picture was of a different woman sitting on his face, skirt around her waist."

"Holy shit! Eight Mexican whores sat on his face?"

"No. There were eleven but only eight pictures in the Polaroid pack."

Les laughed, a deep hearty laugh. Everyone on the bus looked at us. Slapping his knee when he stopped, he asked, "What did he do?"

"He panicked, grabbed the pictures and ran over to the clinic. Not an easy run. It was almost a half mile in his underwear. Seeing you running to the bus half-dressed reminded me of the story."

I continued, noticing the guys in front of us had turned around and everyone within earshot listening. "He ran into the clinic shouting, 'Emergency!' waving the pictures. A medic stopped him before he got too far and tried to calm him down. 'What's the emergency?' he asked. Wurtzberger shouted, 'I ate out eleven Mexican whores last night! I'm going to die.' The medic grabbed the pictures, looked at them, and started laughing. Another medic came to assist him, and Teddy pleaded his case, took the pictures from the first medic, and gave them to the second medic, who started laughing, too. This didn't stop until everyone in the clinic had seen the pictures. When word got around about the trainee in the emergency clinic and the eleven Mexican whores, soldiers from nearby buildings came around to see him and the pictures."

"That's it? They just laughed at him?"

"When two of the medics stopped laughing, they put him in a cubicle. He was still hungover, so they let him sleep for a while."

The guys on the bus must have thought that was the end of the story because they turned away.

"But, that's not the end. Word had spread all over Lackland about this trainee, but not everyone could get over to the clinic. Besides, the medics wanted to have some more fun with him. When he woke up, he found his lips bandaged. Not bandaged shut, but tape covered his lips and a great deal of skin around them. He learned later this was his Scarlet Letter, not so much to shame him, but just to mark him, so all those who hadn't seen him at the clinic would know him by sight. The medics told him to wear the bandages for a week."

"Funny story, but I think you made most of it up—"

"No, it's true," an airman in his work blues said from just behind me. "I heard the story during the VD lecture in Basic at Lackland. It started off with something like, 'A lot of you guys think you're going to stay away from the whores this weekend, but let me tell you a story.' The story was scarier than the VD film."

The story didn't shake Les at all. "I'm very particular about who I sleep with. Like all my ladies, them gals yesterday were sweet little things."

The bus pulling to a stop in front of the South Post mess saved us from listening to the rest of Les' story. I looked at my watch, 0710—lots of time.

The line was short, and our selections were meager, except for Les, who went heavy on the biscuits, gravy, and grits.

There was little conversation during the meal. Not until we finished eating and lit our post-prandial cigarettes did anyone speak, and it was Kennedy telling us the time.

"Seven-thirty," he announced, and we all checked our watches to confirm.

"Let's finish our smokes and head over. Don't know how long it might take to find the office."

No one finished their cigarette, instead opting to leave.

We joined the wave of soldiers, sailors, airmen, and civilians walking through the tunnel under South Washington Boulevard to the Pentagon. Officers and enlisted men mingled in casual conversation as they walked. The main topic was the terrible season the Senators were having, and whether they could survive as a franchise. No one seemed to be a strong supporter of the team, more of a sympathy fan, due to the temporary status of their assignment to Washington. While we listened to the surrounding conversations, the four of us remained silent, overcome by the enormity of the life change we were about to experience.

A sidewalk from the tunnel brought us to the Mall entrance of the Pentagon, the entrance most often photographed. Along with the pedestrian traffic from South Post, a procession of cars dropped off passengers at the entrance. These weren't moms dropping off dads at work, but rather staff cars with Generals. As each car pulled up, a soldier waiting at the bottom of the stairs opened the rear door, saluting the passenger as he got out. The process was repeated three times as we walked the short distance from the tunnel, past the helicopter landing pad, to the front door.

The Mall Entrance is impressive. Set in the middle of the north wall, it looks out over a parking lot to a mall almost two hundred feet deep. The formal entrance is at the top of a short set of stairs, under a portico supported by over a dozen large, square columns. Beneath the portico, massive oak doors open to the interior.

We followed a three-star General up the stairs into the building. To my surprise, as he passed through the door, he extended his hand back to hold the door for us—just a fraction of a second, but it was a gesture of courtesy I didn't expect. Les smiled, opened his mouth to say something, and I whispered, "Don't."

Instead, I said a polite thank you to the General and continued through the door. Les said, "Don't what?"

"You know."

Inside the building, everyone moved with a purpose. I asked a soldier at the door for help finding the office listed on our orders. He directed me to the Information Booth to the side of the Main Entrance, where a civilian lady sat answering questions.

Showing her my orders, with the room number highlighted, she responded before I could ask her the question.

"Take this corridor in front of you to the first cross corridor. That's the beginning of D ring. Go down one flight to the first level—that's '1', the first floor. Turn right and find corridor '7'. Turn left until you find room 726, 1D726." Before I could say thank you, she was helping someone else.

The corridor she directed us to proved to be more an art gallery than a corridor. Paintings aligned both walls. Paintings of soldiers and sailors painted by soldiers and sailors. Some depicted the rigors of war, while others captured more of the soul of the artist, and the valor or futility he tried to convey. One painting caught my eye: an aircraft carrier about three by four feet, which captured the middle of the ship. Airplanes were less than one-half inch, yet I could make out the pilot in the cockpit. The painting fascinated me because of its realness, looking more like a giant photograph than a painting.

At the end of the corridor, we entered a stairwell that brought us to the first floor, turned right and found corridor "7." A left, and we moved down a smaller corridor, with offices to the left and right. When we got to room 1D726, Dexter W. stuck his head in the door and immediately pulled back.

"What?" I asked.

"There's a lot of people in there."

I stepped past Dexter W. and walked into the office. He was right, there were a lot of people, and near the door sat the Command Sergeant Major who rescued us on Friday. I presented myself to him.

"Command Sergeant Major. Reporting as ordered." And presented him a copy of our orders.

He looked up, then looked me up and down. "Nice uniform, West."

"Thank you, sir. I bought it at the PX."

"Don't call me 'sir,' Private. I work for a living," he said with a stern look that changed to a smile as he looked to the two men who sat to his right, a Major and a Lieutenant Colonel.

"Sergeant Major is a little cantankerous this morning," commented the Major, a pleasant man with a southern accent.

"Sir, seeing these soldiers here now reminds me of the poor way they were treated Friday by a young Corporal at North Post."

"Can't have that now, can we?" The Major looked at me, then called out over his desk to a soldier sitting across from him. "Morris. We have replacements to process. Treat them nice, or the Sergeant Major will have your ass for lunch." A smile complemented the words.

The man the Major had addressed rose from his seat, and, with a ruler in his hand, approached me and told me to join the others in the hallway. This first encounter was not pleasant. The man stood a little under six feet tall, greasy dark hair and overweight, his pear-shaped body straining the buttons on his summer-weight Class A greens.

In the hallway, he told us to line up against the wall. Once there, he paced in front of us, slapping his ruler against his leg, mumbling something about the honor bestowed upon us. As he turned to walk back the other way, two soldiers came down the hallway carrying coffee. One, who looked like the television personality Tim Conway, but with a mustache, said, "Looks like Teddy forgot his riding crop today and has to use his ruler."

The other soldier responded, "Ya think? Maybe he forgot his horse, and feels overdressed using the crop. But he still has that strut. Is that his Patton strut, or his Hitler strut?"

"I think that's the Morris strut. When he does the Patton strut, he wears a pink water pistol, and when he does the Hitler strut, he puts a little pencil under his nose."

Morris stopped pacing.

The Conway character asked, "Are you the guys from Fort Dix?"

I responded, "Yes."

"I've got your assignments. Come with me."

And we did, entering an office that was a different world, leaving Teddy in the hallway.

CHAPTER 14

"Hello Andy"

We followed the Tim Conway character into the office, past the Command Sergeant Major and the two officers, none of whom gave us a glance. The room was big, about the size of three grammar school classrooms, L-shaped, two rooms deep, with the third off to the left. Desks fought with each other for space, often four across, touching the desk on either side and touching another row of four which faced them. The only desks having any space around them were those of the two officers in front. A quick look around and I estimated there were ten soldiers mixed in with twenty civilians, all seeming to work together.

Halfway into the room, he turned to the area on the left and took a seat behind a desk facing only one other desk. Once seated, he introduced himself.

"Gentlemen, I'm Andy Henderson. Call me Andy. The only ones in this office you have to address by rank are the Major; the Lieutenant Colonel, who you'll call Colonel; and the Command Sergeant Major, who likes being called Sergeant Major. The asshole in the hallway likes to think of himself as the Spec-Five of the Army, but no such rank exists, so we don't call him anything. We never use his rank and try not to use his first name. Who's got a copy of the orders?"

Each of us offered him a set, but he only took mine, saying, "You all came here on the same set, so I only need one. Before you leave here today, you'll each get new orders."

Looking at the orders, then to a document on his desk, he asked, "Who's Kennedy?"

Jimmy stepped forward.

Henderson smiled. "You had some people going, you know. Kennedy from Massachusetts got a Major up at Fort Dix all excited. He made enough noise you got assigned to the Joint Chief of Staff's Office before I realized you pulled one over on him. For that move alone, you deserve JCS, and that's a good thing." With that, Henderson rolled a piece of paper with five colored carbons into his IBM Selectric and typed. When finished, he got up went over to the Major for a signature. Returning, he gave the yellow copy to Kennedy. "Take this to 2D932, your new home. They're expecting you."

Dexter W. was assigned to the Senior Enlisted Branch somewhere else in the building, and Les went to Combat Arms, which was four offices down the hall from where we stood.

"West?" Henderson asked.

"Yes."

"Sit down," he said pointing to the desk abutting his. "They brought you down to work at Walter Reed Army Institute of Virology on a new strain of syphilis that turned up in Vietnam. Probably because of your undergraduate degree in biology at Harvard."

I nodded. Pleased someone had selected me for such an important assignment, scared when I heard myself referenced in the same sentence as Vietnam and syphilis.

"The assignment requires another level of security beyond the Secret one you have, so there will be a wait while they go through the process. In the meantime, you'll stay here working with me on temporary duty until it comes through."

I must have looked perplexed because he added, "It's no big deal. There's always a TDY here waiting for security. Mac over there"—he pointed to a blond guy across the room—"is doing temporary duty waiting for a Top Secret. Then he can join the Courier Service, transporting documents back and forth between Washington and Europe. Sounds glamorous, but it's not."

"What will I be doing for you?"

"Typing, filing, answering the phone, maybe some special assignments, research-type work. Most of it'll be boring, but there might be a few things you find interesting. You can use this desk this week and next. Frank, my boss, is on two weeks' leave. He's ETS-ing next month, so he's looking for an apartment in Philly, where he has a job."

"ETS-ing?"

"Getting out of the Army. Expiration Term of Service."

And that was it. Andy put me to work right away, first with some filing, then typing form letters. He introduced me to a few of the people in the office, but they were less interested in knowing me than I was in meeting them. I wouldn't be here that long.

Andy took me to lunch with him and the guy he was with in the hallway this morning. A guy named Pat Mullaney. Mullaney and Andy had a bond of being law students. Pat was a first-year student in Nebraska when he got drafted, and Andy had been in his second year at Columbia. Both had been offered direct appointments as officers in the Judge Advocate General Corp, and had refused. Accepting the appointment would have meant a three-year military obligation, instead of the two they had as draftees.

The two had an easy banter between them, joking and laughing, sharing the same sense of humor. They were a contrast physically. Andy was almost completely bald, a condition he says happened in the short time he was in Fort Polk, Louisiana, and while not fat, looked soft. Mullaney, on the other hand, looked fit and had a head of thick, dark hair.

They quizzed me about my background, both interested in life at Harvard, but neither asked nor seemed to care about my brief stint as a professional hockey player, or my selection as an All-American. I guess they were more the brainy type.

After work, Les and the others were waiting for me in the barracks. Kennedy had been back to the room he shared with Dexter W. during the day and cleared his stuff out with not even a note. We changed into our civilian clothes and spent an extended dinner hour talking about our days.

The guys were excited about their assignments. No one had a clear idea of what they'd be doing, just a general sense it was clerical. We all had normal hours, eight to five, Monday through Friday, no extra soldier duty, and no weekends. Both Dexter W. and Les had been told they would stay where they were for the remainder of their time in the Army. Just a bad-paying job, but they could make plans.

Dexter W. had called his wife, and she was excited he had a permanent spot in Washington. She would quit her job and move to

Washington. She was coming down the following week to look for a job and apartment.

Les was single, so we asked about a girlfriend and he smiled, reminding us he already had two girlfriends from the weekend. Right, how could I forget? He already had a line on an apartment complex near Washington National Airport, where a lot of Mohawk and Allegheny Airlines stewardesses had a home base. Sounded like heaven to him.

That left only me. I told them about my day and my TDY assignment. Les was quick to point out it wasn't an assignment, but rather a lack of assignment. "That sucks!" he said. "You could be in limbo for months and find out you didn't get the clearance. What happens then?"

I told them how I would work with Andy while I waiting for my security clearance. Worst case, he'd find something else for me.

"Sounds like staying right there would be the best case," Les said, sounding sincere and non-corn pone, which surprised Dexter W., who had heard nothing but the corn pone. "I ain't no scientist, but it seems to me if you got to study this super syph bug that's making the guys in Vietnam sick, then you got to think there's a strong possibility they'll send you to Vietnam for a close-up look."

Dexter W. agreed and sighed, "Make's sense," and it did make sense. Getting the added security clearance was not the best case; it could be the worst case if it included a trip to Vietnam.

Les looked at me thoughtfully for a long minute before saying, "Well, little buddy. Seems we have some work to do. All of us, including you, have to see what we can find out about guys who study the super-syph bug. Do they go to Vietnam? Do they stay in Vietnam? Might even be good to know if anybody knows how safe it is to work with that bug."

"How the fuck are we supposed to find that out?" I asked.

Les looked at me and said, "Guys, this ain't rocket surgery."

Dexter W. interrupted Les. "Did you mean brain surgery or rocket science?"

Annoyed at the interruption, Les asked, "What did I say?"

"You said rocket surgery."

"Glad you were paying close attention, 'cuz that's what I meant." He continued, "An Army travels on its stomach, shoots with its guns, and thinks with its dick. Every guy in Vietnam is thinking about pussy, and the only pussy around is Vietnamese pussy. So, if they're all thinking about it, they're all getting it. If some of them are getting super-syph over there, they're all thinking and talking about super-syph, especially the lifers. And gentlemen, one of us is working with the senior enlisted soldiers, which handles all the lifers. And you, good buddy, have a really good senior enlisted in your temporary office, the Command Sergeant Major. So, all we gotta do is ask and listen."

"Les, I've only known you for a few days. I'd like to think we're friends, but we're not the best of friends. So, why do you have this interest in my future?"

Les smiled and went corn pone for the first time in this conversation, "Gomer, I do like you. But," looking over his should in a conspiratorial way, "unlike you, I learned something important today," and he thumped me on the shoulder.

"What?"

"I asked about us processing through 1D726. That office we came through today is EPCMR-GS, or Enlisted Personnel Career

Management and Retention, General Support. It handles every Army assignment in the world, and has the IBM card for every swinging dick who's an E-6 and below with a seven, eight, nine, or zero series MOS. We are E-2 privates with a seven series MOS. That office controls us for the next two years. Now do you understand?"

"If West stays in that office, he controls our IBM cards! He can let us know if we're getting reassigned," Dexter W. added.

"Wrong," I said to everyone's surprise. "I learned today everyone in that office pulls their own IBM card every month when the new batches are printed. They have their IBM cards and those of their friends under their desk blotters. I won't be able to tell you you're being reassigned, because you can't be reassigned. With your IBM cards under my blotter, you disappear. The Army doesn't know you exist." I looked at Les and he was beaming.

"I told you it wasn't rocket surgery!"

"Then it's settled. Tomorrow I'll go in and ask to stay in that office. No need to wait three months."

"Sorry, little buddy, but we got too much riding on this, all of us, including you," said Les, taking on the leadership role in this conspiracy of conscripts. "There ain't no guarantee you can stay in that office after just one day, or one week, or a month. You said it—they expect you to work there temporary until your clearance comes through, and if it don't, and if you been doing a good job, and if there's an opening, they might ask you to stay. Seems like you gotta take it slow, but more important, you gotta do a good job."

"Besides," added Dexter W., "It'll give us time to find out about this super-syph bug and the job at Walter Reed. From what I hear, we're all safe in our jobs, so the chances of us being reassigned are remote—only if we fuck up." Then he said, "Let's make sure everyone gets a fair deal," as he looked at me with emphasis. "This Walter Reed thing, if it's safe, could be a career-maker for you. You could meet some really important scientists, and be on the inside to a great future. Hate to see you have regrets about what could be a great career, just to protect our asses."

Dexter W. was right, and I agreed with him saying, "Yeh, let's get all the information, cover all our bases, and make a decision later, a decision that's good for all of us."

"Okay." Les took over the meeting again. "We have to find out what we can about the super-syph job, and the senior NCOs who work at Walter Reed on the project. You"—he said, pointing at me—"have to make yourself in-fuckin'-dispensable to the running of EPCMR-1D726, or whatever the fuck it is. You also have to find out about vacancies coming up in that office. We don't want them to go looking for someone else, and get blindsided."

"That's enough for now. Let's get dessert. We got all the big details, and we can work out the small details later."

Before we finished dessert, we decided we had enough to do without over-strategizing. We would collect what information we could for a week, and discuss it next Monday night at dinner.

CHAPTER 15

"Settling in and heading out"

I spent the rest of that first week typing for Henderson. He said it was the best way to learn about everything going on, and he was right.

Henderson was responsible for a lot of things that required special handling, things needing careful thought. One of his jobs required him to select soldiers who wanted to work in an Armed Forces Entrance and Examination Station, or AFEES, a central processing facility for all the branches of the military. Like everyone else who'd been drafted, it had been my first exposure to the military. When I got my draft notice, I went to an AFEES for my physical, and I reported to an AFEES for induction. To handle these functions, the Army needed permanent staff stationed there, mostly clerks, typists, and medics. EPCMR-GS handled all these jobs. These were good, safe jobs, and Henderson took care in making the selections. He tried to assign guys near their hometowns, so they could live with Mom if they wanted. Soldiers stationed at an AFEES had no KP or guard duty. Pretty soft duty.

Henderson filled the job requests for recruiters. Recruiters were lifers. Most had been to Vietnam and had been to the Army Recruiter School.

Henderson also had to fill MAAG, or Mission, assignments. This meant providing soldiers for our embassies around the world. Even with over one hundred embassies, these jobs were a premium, because Marines filled most of the positions.

All this I learned from asking Henderson questions, while I typed his response to soldiers who had applied for one of these jobs. Most of them got an NFC, which was Army shorthand for "Not Favorably Considered," or in some cases "No Fucking Chance." These form letters had a prescribed format of numbered paragraphs. The first paragraph, numbered 1, stated only "Favorable" or "Not Favorably Considered." The second paragraph of a "Favorable" letter was a brief description of the position for which the soldier applied, and subsequent paragraphs gave specifics of what the soldier had to do. The second paragraph of the "NFC" letter contained a short description of why the soldier didn't get the assignment. Henderson spent a lot of time writing these second paragraphs for the NFC letters, and had a nice way of saying "You are too fucking stupid" that made it sound like a compliment. He would do well as a lawyer, the way he manipulated words.

I didn't learn much about the super-syph job at Walter Reed from the IBM assignment card in the office file. None of the eighty-eight coded characters gave a clue about trips to Vietnam.

EPCMR-GS was the clearing house for all the junior enlisted Army jobs in the Pentagon, the Military District of Washington, which included Fort McNair, Fort Myer, and several military and government offices in and around the District of Columbia. Agencies such as the 902nd Military Intelligence—which had responsibility for the internal security of the Pentagon, National Security Agency, and the Military Courier Service—all sent their special requests to EPCMR-GS.

Each day presented a learning experience. At least twice a day, I'd get a letter to type that was out of the ordinary. I'd ask Henderson for the background behind the letter. He'd always take the time to answer my question with patience, and always finish by asking me if I understood. I'd shake my head yes, but silently say, "Holy shit, I can't believe what's going on in this little office."

I also learned how the jobs in 1D726 got filled, or how I got here. The Major liked to get a new guy on board about six months before the guy being replaced was leaving. How did they get these people? Turns out it was exactly like they got me and the other guys who came down from Fort Dix. They'd wait to get requisitions for multiple jobs around the Pentagon or MDW, and ask Fort Dix to screen guys in the clerk school. When EPCMR-GS thought they might need someone, they'd just pad the request with one or two extra guys. They'd look at everyone who came in, and keep the best for themselves. If no one fit the bill, they'd put the extras to work in 1D726 until a new request came in. Then, they'd get bonus points, because they'd be able to fill the request immediately, and emphasize they were giving up one of their own.

I got to meet other guys in the office, and continued going to lunch with Henderson and Mullaney. I don't think the law school bond was as important to Henderson as it was to Mullaney. Joining them for lunch flattered me, until I learned they felt I lacked the experience to leave me at the desk by myself. Seems if someone occupied the desk, it was expected someone could provide service. If I was there by myself, chances are the service or information would be wrong, and Henderson would have to fix it later. So, my inclusion in their lunch bunch was nothing more than a time management thing for Henderson.

Mullaney seemed a nice enough guy. The physical contrast to Henderson was striking. While Henderson had gone prematurely bald,

Mullaney was semi-simian, having hair everywhere. His arms looked like he was wearing a fur coat under his short-sleeve shirt. Even his knuckles sprouted pads of hair. By noon, he needed a shave, something officers outside the office pointed out. But his most prominent hairy feature had to be his eyebrow, not eyebrows. It emerged from one ear, ran across his forehead and disappeared into his other ear. He relished this feature. While the eyebrow at the temple area was thin, he had a way of scrunching his face to emphasize the hair there that made the ear-to-ear eyebrow a reality. When he did it, he smiled at the "What the fuck?" look he got.

My lunches with Henderson and Mullaney taught me one thing. These guys were smart! Henderson had gone to undergraduate and law school in New York City before he got drafted. During those years, he became addicted to the New York Times. He still read it every day, from front to back. Reading the Sunday New York Times became a religion to him. He and his wife gave it the devotion it required on Sunday, but extended this attention to this bible throughout the week, saving certain sections for certain days. When they were in New York City, they left the "Arts and Leisure" section for Thursday, to plan their weekend activities. The habit remained when they came to Washington, even though it was useless in planning their D.C. activities. Mullaney read the Washington Post every day with the same zeal Henderson dedicated to the Times.

Both loved politics and besides current events, paid a great deal of attention to what was going on in Congress, located only a few miles from where we were having lunch. The Vietnam conflict was always a topic with them, alternating positions as "hawk" or "dove" as it pleased them. But what really fascinated me was the wider range of their conversations. They discussed books they were reading, the arts, and theater—not movies, but real theater. This was totally foreign to me. I played hockey before I got drafted. Before that, I was a science undergraduate. There, discussions veered toward the classes we were taking, sports, or girls. In retrospect, we weren't much in tune with the real world, and worse off for it. Now, I was getting schooled in a world I didn't know existed, and loving every minute.

As the week moved toward Friday, everyone talked about the plans for the weekend. Les met a stewardess, and got an invitation to join her on a trip to Las Vegas for the weekend to see an Elvis performance. Airline employees, including stewardesses, got ridiculously low prices on domestic air flights for themselves and family. Les would be her brother for the weekend. Miss Stewardess must have been something for Les to agree to go to Las Vegas to see Elvis. Ever since the Beatles had shown up, his popularity had taken a nose dive. This must be the Elvis farewell tour. Keep the ticket, Les—Elvis's last concert might be a collector's item someday!

In the office, there was less talk of travel, as most of the guys had been there for a while and had settled in. I told Henderson I'd like a pass for the weekend to head home.

"You don't need a pass unless you plan on flying, and then only if you want to fly military standby. But, with all the military heading out of Washington every Friday, you might not get on a plane until Saturday. As a Private, you have a low priority," he said. "Where are you heading?"

"Upstate New York, near Albany."

"How do you get there from here—train, bus?" he asked.

"Don't know. This will be my first trip since I got here. Any suggestions?"

Henderson looked around the office. "Morris goes home every weekend to Wilmington, Delaware. Takes the train that continues north. He can tell you the schedule, but beyond that, you'd have to check the schedule up to Albany. Wait a minute. Vinnie lives in northern Jersey, he might know more." He called across two desks to Vinnie Rinalli. "Hey Vinnie, West wants to get to Albany this weekend. Any ideas?"

"He's got two choices, the bus or the train. With the train, he's got to get into the city. Take the train from here to Pennsy station, then get across town to Grand Central to go upstate. With the buses, he's got more choices, and it's cheaper. I'm drivng home this weekend to Saddle River. There's a Greyhound station in Nyack where he can get a bus to Albany. If he checks with the bus station, and the timing works, he can ride up to Nyack with me, and get the bus from there. I can't make any promises for a ride back, though."

On the phone, I confirmed a bus station in Nyack, but it was for Trailways. There were two buses from Nyack to Albany, one at eight p.m., and the last one at ten p.m. I tried to make a reservation, and the lady told me they didn't take reservations. I got the address of the bus station and told Vinnie. He said he knew where it was.

"Hey, Andy," he called out, "do you think there's any chance we can work Teddy's schedule on Friday, and get a jump on the traffic?"

"Makes sense, but I don't make the rules. Have to clear it with the Sergeant Major. I wouldn't use the Teddy argument though. He doesn't like to feel he's being put in a corner. Just be up front with him."

Vinnie came over to my desk and said, "Let's go see what Stoner has to say."

Stoner didn't even look up from his desk to see who was asking. He said yes before Vinnie could finish the question. We said our thanks and went back to work.

Vinnie said he was convinced Stoner had dog's ears, and had probably heard our conversation before we got over there. I looked over at Stoner, and he was looking our way. He pointed to his ear and winked at me. I vowed to be careful what I said above a whisper in this office and vowed then that anything I said about Stoner, even in a whisper, would be complimentary. A quick glance at Stoner again and he was still looking at me, only this time he tapped his forehead and winked. That old fucker can read minds, but then quickly corrected the thought to: What a remarkable man.

Stoner smiled!

During the week, I learned of even more specialized functions taking place in the office. Other people assigned soldiers to special advanced training schools, such as the Defense Language Institute, or DLI. The office also made the cushy assignments for athletes, such as golfers, who got to spend their tour of duty teaching golf to Generals' wives in Southern California. One desk only handled reenlistments from soldiers who, while in Vietnam, decided they wanted to reenlist or extend their tour of duty in Vietnam.

"That must be the easiest job in the office," I commented. "Who would want to reenlist when you're in Vietnam?"

"You'd be surprised. Remember, most of the guys we handle are not on the front line. They're in back areas that don't see much enemy action, and the duty can be pretty good. So, if they're not in any danger, and they want to get a few more months of combat pay before they get out, they ask to stay. Some other guys looked up the Army regulations. If they come back to the States with less than one hundred eighty days before they get out, the Army is obligated to let them out for good, as soon as they hit U.S. soil. So, these guys figure out how much time they've got left, how long it will take to process out of Vietnam, and fly home to Oakland or Fort Lewis. Then they set their target at a day or two shy of that. They ask for an extension of their tour to coincide with that date. It's pretty smart if you have a safe assignment," Henderson explained.

"What about those guys who reenlist while they're in Vietnam? That doesn't sound smart," I asked.

"Part of it is smart. Most reenlist after they get a week of R and R in Japan or Hawaii. After they spend time with their wives or girlfriends in the real world, and have to go back to Vietnam, they hate it more than ever. They reenlist to get a long tour of three years. They might have another stripe after being in Vietnam, and ask for Germany, where their wives or girlfriends can join them. There's some risk, because when they finish their tour in Germany, they'll still have over a year to serve, and can be sent back to the 'Nam. They're betting the war will be over by then."

Henderson continued. "There were a lot of reenlistments after the Tet Offensive. Guys just wanted out of Vietnam. Guys who had already extended their tours got fooled, thinking they were safe. The Tet Offensive left no safe areas. It's still a crap shoot, but they know the odds. Our job is to help as many of them do what they think is best for them. At least, that's the way I see it."

The rest of the week went by without incident, except on Wednesday, July 30, they announced President Nixon made an unscheduled trip to Vietnam to discuss Vietnamization. Such an agreement would reduce the U.S. troop commitment, and shift the emphasis of our office to bringing soldiers home.

I called home to tell my parents I'd be up for the weekend. They were excited, but I still harbored anxiety about the unsettled nature of my future. When would I know when or if I'd go to Walter Reed? If I went to Walter Reed, would I have to go to Vietnam? Could I stay in 1D726?

And, the big question everyone had, if I went to Vietnam, would I come home?

After lunch on Thursday, as Henderson and I walked back to the office by ourselves, I screwed up the courage to tell him about my reservations about the super-syph assignment, and planted the seed about the possibility of staying in 1D726.

"Well, you match up pretty well against the requirements for my opening, so it's all about your ability." Then he added, without prompting: "And from what I've seen in these three and a half days, I hope you don't get the clearance. You seem to fit in. Let me talk to the Major about you. Frank gets out in two weeks, and I'm his replacement. Then I have to train my replacement. It could work out for you."

"Thanks," was all I could mutter.

"Oh. Stoner said to tell you he likes what you've been thinking."

I stopped. "The Command Sergeant Major can read minds?"

He laughed, the most genuine laugh I'd seen from him to date. "No, but he has a great sense of humor when you get to know him."

Friday morning, I showed up for work, ready to travel, needing only my shaving kit. This was in stark contrast to Teddy, who showed up carrying a stuffed duffle bag.

"Where's he going?" I asked Henderson.

"Teddy goes home every weekend to Mom and Dad, mostly Mom, so she can cook and do his laundry. That's what's in the duffle bag—dirty clothes, including bed linen and towels. We guess Mom and Dad meet Teddy at the train station, and take him to dinner at his favorite hangout, probably the one he hung around as a teenager, eating fries and drinking cherry cokes. He gets a chance to tell Mom and Dad, and anyone within listening distance, how he's giving all the Generals at the Pentagon advice on how to win the war. Then they go home where he can watch TV until the test pattern comes on. Probably sleeps 'til noon on Saturday, while Mom is doing his laundry and making his favorite meal for Saturday night. Ditto for Sunday. He gets the early train on Monday morning, and gets into the office about thirty minutes late, with a duffle bag full of clean clothes."

Henderson said all this without emotion. He just recited the facts as they had been fabricated by the other guys in the office. No one had ever bothered to ask Teddy what he did at home. They preferred their version. It fit with their view of him as a lazy asshole.

"But if he's such an asshole, why does he have the top job in the office under the Command Sergeant Major?" I asked.

"He thinks he has the number two job in the office, but he doesn't. The Major brought Teddy in to replace Al."

"Who's Al?" I asked.

"Al is the number two enlisted man in the office, after the Sergeant Major. Al's on leave now, so you haven't met him. When Teddy started learning Al's job, the Colonel, the Major, and the Sergeant Major realized he couldn't do it, and was an asshole, but they couldn't do anything about it. We couldn't give him to anyone else because he'd be too dangerous, and it would hurt our image if we assigned someone like him. So, we kept him. He thinks he's learning Al's job, but he's not. They moved a lot of Al's stuff to my desk. When Al leaves in about sixty days, and Ray starts the learning process under Teddy, we'll feed the stuff back to Ray. Teddy will never know what he's not doing, because Teddy has never known what he's doing, or not doing," Henderson explained, in the same matter-of-fact way he had explained the duffle bag.

"Un-fucking believable," was all I could manage to say.

"No, it's the Army way," corrected Henderson, "even at the Pentagon."

"Who's Ray?" I asked.

"Ray's the office stenographer, a history major from the University of West Virginia."

"Have I met him?"

"No, he's on leave."

"Is everyone on leave?"

"No, I'm here. I don't take leave in August. Lots of people do, but not me."

On Friday afternoon, the clock seemed to go into slow motion.

Anxious to get underway, I kept looking over at Vinnie, seeing if he was getting ready to clear his desk and head out, but he gave no early signs. I started winding down about two-thirty, racing through a letter, and then taking my time to put the documents in order. I rearranged the piles on my desk. Finally, at about ten minutes to three, Henderson leaned over to Vinnie and said, "Vinnie, wrap it up and get him on the road. I can't stand watching him rearrange those papers on his desk any longer. He looks like a poor imitation of Teddy trying to look like he's doing something."

Vinnie looked up at Henderson, and then at me, and said, "I'm ready. Let's went." He cleared his desk before I could.

As I put the last of the typed documents in front of Henderson for review, I said, "Thanks."

Vinnie and I were out the door and on our way. Vinnie was off to a normal weekend with his family and friends in Northern Jersey, and me, unknown at the time, off to become a peripheral observer of a historic piece of Americana.

CHAPTER 16

"History in the making"

Vinnie had parked near the South Post Mess Hall, so we got to his car quickly. And what a car—a Triumph TR250. Dexter W.'s Chevelle was nice, but this was more my style. Vinnie hurried to beat the early Washington traffic. But he wasn't in such a hurry he didn't have time to put the top down. The task was a one-man job, made easier and quicker with two sets of hands.

With the top down, we jumped in. He hit the ignition, dropped into first gear, and spun the rear tires on the dirt and gravel before I could appreciate the throaty rumble of the sweet little product of British automotive craftsmanship. He kept the car in second gear until we cleared the South Gate and headed onto the Expressway. Then he made a full run through the gears and a full run through the traffic, until we reached the Baltimore-Washington Parkway. Once on the Parkway, he relaxed and settled into his entitled position in the outside lane.

Vinnie's entitled attitude came naturally, and extended beyond the driver's seat of a sports car. He had a country-club air to him which he wore easily. A little under six feet tall, he had the classical dark Italian features: dark hair with a wave in it, made visible by the longer-than-regulation length; dark eyes that could be both piercing and playful; and olive skin made even darker by time in the sun. All his tan couldn't have come from driving the TR250. Some must have come from golf, tennis, or the country club pool. I don't think he was headed home to work in the family's truck farm. The only visible flaw to his Italian prince persona was a gap between his two front teeth that had escaped the high-priced hands of an orthodontist.

With the top down, and the speed up around seventy, we couldn't have much of a conversation. There was a little traffic around Baltimore, but Vinnie maneuvered the TR250 into holes in adjacent lanes and always had the advantage. When we stopped for gas at the rest stop south of Wilmington, Delaware, he told me, "I could tell you I'd let you drive, but that would be a lie. The fact is, nobody drives this car but me." Not a problem for me, just get me there in one piece and on time.

I watched Vinnie when he shifted with a forward throw to first or third gear. The forward throw on a TR250 was too long. If you wrapped your hand around the shifter on the forward throw, your knuckles hit the push buttons on the radio and changed the station. When the Brits designed the car, they didn't include a radio. The engineers assumed the sound of the engine and the exhaust provided all the music the driver would need. The U.S. market demanded a radio, and the knuckle problem came with it. Vinnie took the forward throw with the palm of his hand doing all the work. The pull from that position was accomplished with a cup of the hand from above on the top of the shifter.

Vinnie was a good driver and knew this car.

At Wilmington, we crossed the Delaware River and got on the New Jersey Turnpike, with Vinnie continuing his preferred position, racing up the outside lane. The only words between us were when we passed the Freehold exit. He said, "We're making good time. Piece of cake to make the early bus."

My answer was lost to him. I couldn't even hear it myself as he blew by a loud eighteen-wheeler. When we exited the Turnpike for the Garden State Parkway, the traffic heading south stood still. This was the evening rush hour for the people who worked in Manhattan, heading home to the suburbs of New Jersey. Being August 15, a lot of the traffic was headed to the Jersey Shore for the weekend. By stark contrast, the traffic heading north and away from the shore was light. Vinnie moved into the center lane and held a steady seventy, slowing only for the toll booths. We got off the Parkway, and he wound his way through neighborhood streets, finally emerging on route 9W, just south of North Nyack.

He did a mini-power slide into a mom-and-pop grocery store/gas station a little after seven-thirty proclaiming, "Pretty good time, considering," presumably to me, but I wouldn't know what good time was. Vinnie flew the whole way, so I gave him an affirmative nod. He said I could get my ticket in the store. I started to thank him, and he shrugged it off.

"Sorry I can't offer you a ride back, because I don't know when I'll be heading back. If things don't work out with the girlfriend, I'll head back tomorrow night. If things go well, I might not leave here until late Sunday night."

"Hey, no problem. I understand, and appreciate the lift. Probably saved me a couple of hours. I'll be home by ten-thirty."

"Glad to help," he said as he eased the clutch out. "See you back on campus on Monday." As he accelerated in first gear, I got another chance to hear the deep rumble of the TR250. What a sweet car!

I turned and went into the grocery store/gas station. Mom and Pop stood behind the counter, looking like something out of a Norman

Rockwell painting, if Norman Rockwell had ever thought of painting a picture of a mom-and-pop grocery store/gas station in New Jersey. I nodded a polite hello and said, "I'd like to buy a ticket on the next bus to Albany, sir."

"The next bus should be along here in about twenty minutes, son. That'll be twelve dollars for a one-way ticket, and twenty for the round trip. The return ticket is open, so you can come back any time you want," the pop figure said in a practiced monologue. He added, "Sorry, but the bus company don't give no discount for soldiers, like the airlines do."

"That's okay sir," I replied. "I'll just need a one-way ticket," putting a ten and two singles on the counter. "Do you have a pay phone?"

"Just one way?" he asked.

"I just got assigned to the Pentagon," I bragged, "and I'm going home to get my car," I lied.

Pointing to the door, he told me the pay phone was on the side of the building. At the phone, I dropped a coin in the slot and dialed the operator.

"Operator, may I help you?"

"Yes, operator. I'd like to make a collect call to 555-123-5188."

"Who may I say is making this collect call?" she asked.

"Collect call from Jonathan," I responded, giving her the prearranged code word, Jonathan, which meant I was on the early bus from New York. This way my father would get the information and wouldn't have to pay for the collect call. If it had been the later bus, I would have said Jon.

I heard my father answer the phone, and the operator go through her message. I could also hear my father when he said, "No, operator. There's nobody here right now. If he could call back at ten-thirty, there might be someone to accept the call," he said, with the call back time being the acknowledgement he would meet the bus at ten-thirty. The operator passed along the information, even though I'm sure she knew I heard it all, just as I knew she knew my father and I had just had a conversation without having to pay Ma Bell.

As I stood outside waiting for the bus, I watched Mom and Pop bring things inside. It looked like closing time. I asked, and they confirmed they opened at six a.m. and closed at eight p.m. every week day, to accommodate the commuter traffic.

"How would I have been able to buy a ticket if I got here after you closed?"

"Buy it on the bus for eleven dollars, just like everyone else," he replied.

"But you charged me twelve dollars," I protested.

"Commission," he said as he locked the front door.

I waited on the steps of the store watching the traffic on 9W. Watching traffic in a strange city isn't as interesting as it had been back home as a kid. Here, there was nobody to wave at. Back home, if I saw Mrs. Berg go by at eight p.m. on Friday night, I knew she was coming home from the Novena service at church. If I saw Marty from the pharmacy go by at eight p.m. on a Friday night, he was making the last delivery before the store closed for the night. But here, they were just cars.

The bus pulled in a little after eight p.m. While I thought I was the only one getting on the bus, it surprised me to hear car doors slamming and people walking toward the waiting bus with small pieces of luggage. They had been waiting in their cars. Having been raised a gentleman, I repressed my urge to be the first one on the bus, and let the women, children, and older people get on before me. This line of people included one woman, and an older gentleman with a child. As I waited, each announced their destination to the driver, and he replied with the price. From the cheaper price, it sounded like they were going a shorter distance. When I got on, I presented my ticket to the driver.

"Don't see many tickets on this stretch of the trip, soldier. Did you buy it from the Parkers?" he asked as he took my ticket.

"I bought it in the store, sir."

"Did you get the military discount?" he asked.

"No sir, full fare of twelve dollars. The man said the bus company doesn't offer a discount to the military."

"So, you paid full fare. If you buy the ticket on the bus from me, it's nine bucks, which includes a one-dollar military discount," he replied with a sympathetic smile. "Least I can do is give you a first-class seat for that price," he said as he cleared his jacket, lunch box, and hat from the first seat on the right-hand side of the bus.

"Thanks," I said, taking my seat as the bus left the parking lot with a lot less pizzazz than Vinnie's TR250. As I settled in, thinking about the Parkers, I concluded, "You can take the people out of New York City, but you can't take the New York City out of some people."

The bus continued north on 9W to Newburgh, stopping every few miles to let someone off. Nobody else got on. After Newburgh, we headed west to connect with the New York State Thruway. From here, I would be in Albany in less than two hours. The rhythmic roll of the bus at sixty miles an hour on the Thruway was more comfortable than the stop-and-go travel on 9W. I aettled back in my seat and let the bus and the road rock me to sleep.

Funny how the body can pick up subtle differences in its surroundings while asleep. I woke up from a pleasant dream, where I was eating a big bowl of strawberry shortcake, made with fresh baked biscuits, fresh strawberries, a little vanilla ice cream, and a lot of whipped cream. I had just tasted my first mouthful when I was awakened. Looking at my watch, I realized I had slept for less than thirty minutes. The bus was still on the Thruway, but slowing—not much, maybe down to fifty miles per hour. From my high-priced seat at the front of the bus, I saw traffic backing up. A big Thruway sign coming up on the right announced we were two miles from Exit 19, Kingston. We were still moving, but slowing all the time. By the time we got to the sign announcing Exit 19 was only one mile away, our speed had dropped to less than thirty miles per hour. I saw three lanes of traffic on this two-lane highway, and the flashing lights of police or emergency vehicles up ahead.

By the time we got to the point where we could see Exit 19, the bus had stopped. There were now four lanes of traffic: the original three that had started further south, but now a fourth lane, which comprised vehicles parked on the shoulder.

"Looks like a bad accident," I said out loud, but directed at the driver.

"I wish," he responded. "An accident they'd be able to clear in a few hours. The dispatcher warned me about this before I left New York. There's a big hippie convention in Woodstock this weekend. You know, loud music, drugs, and sex."

"Is this Woodstock?"

"Hell, no. Woodstock's about five miles west of here, but they got traffic backed up all the way, and in both directions on the Thruway and 9W. The damn hippies are leaving their flower-mobiles on the highway and walking in, adding to the traffic problem."

"Sounds like they're going to be late for the party if they have to walk five miles," I said.

"Hell, no. The damn party is supposed to be all weekend. I've got this run back and forth to Albany all weekend. It'll take me forever."

"Must be one helluva big hall they're having it in. Is there a college or something over there in Woodstock?"

"Hell, no! They're having their party on some farmer's field. They're crazy. It's supposed to rain all weekend."

"Sounds like they'll have a tough time smoking their dope in the rain," I teased with a smile.

The driver teased back, "Sounds like there's going to be a lot of naked girls running around with mud up their cracks."

As we sat in the Thruway parking lot, I decided to take the train back to Washington on Sunday. I had been considering the bus purely based on cost, but as we sat, I realized the extra cost of the train was worth it. I could avoid the traffic sure to be here on Sunday. And, I could avoid the possibility of having to tell my new temporary bosses at the Pentagon I was late because of a traffic jam caused by commie, pinko, Vietnam-War-protesting hippies having a party.

I tried to sleep again as best you can in a stationary bus, and grabbed another twenty minutes of shut-eye before the bus started moving. It wasn't moving directly forward. A State Trooper at the front of our lane was forcing those ahead of us to move more to the right. There wasn't much room to the right. So, to get the traffic going north, he forced the left-hand lane of traffic onto the grass median. Now we were in four lanes of traffic. This was cautious going, but we got through after another twenty minutes. As we accelerated back up to Thruway speed, I looked at my watch and noted we should have been in Albany already. Sure glad I didn't have to take the later bus, I thought, as I drifted off for another nap.

I awoke when the bus slowed for the exit in Albany, recognizing my surroundings even though they looked different from the high seat of a bus, compared to the seat of a car. Not that there was much to see at midnight. The bus wound through the streets of Albany and pulled into the Trailways Terminal at twelve-fifteen a.m., almost two hours late. I saw my dad pacing inside, behind the glass window. A smile crossed my face as I thought of the possibility my dad had retreated inside after being approached by a hooker.

I was the first one off the bus because of my "first-class ticket" purchased from Ma and Pa Parker. My dad pushed through the door, leaving the relative security of the terminal for the land of the hookers.

"Hi, Dad."

"I thought your message said you were getting the early bus?" he asked.

"Hi, Dad," I repeated.

"Hi, son. I thought you said you would get the early bus?" he repeated.

"I did. The bus got caught in traffic near Kingston. Hippies going to a party in Woodstock."

"Woodstock's a good five miles from that exit."

"Yeh. It was a mess. I don't know much about it. Let's read about it in the newspapers tomorrow."

"I called your mother twice since I got here. She'd be worried if we didn't get home on time, and I thought there was a problem along the way, and you might have called to let us know. When she told me you hadn't called, it made her worry maybe you were in a bus accident."

"Why don't you call her now and tell her we'll be home in thirty minutes," I said. I would have done it, but I would have had to listen to how worried she'd been, and then explain everything I knew and everything I didn't know about the traffic on the Thruway. If I had to do all that, I'd miss breakfast. As it was, it took my dad a full minute to say, "He just got off the bus, and we'll be home in a half-hour."

When we got in the car, I said, "Boy, it's a good thing I'm not going to Vietnam. If I missed sending a letter one day, she'd be calling Nixon the next day." He missed the point for a few minutes. All he did was acknowledge she worried a lot about me ever since I got lost as a four-year-old in Hampton Beach. Then it hit him.

"You don't have to go to Vietnam?"

"Probably not. I think I can stay at the Pentagon for the whole time."

"That's wonderful. Let's stop and call your mother."

"No, it's a long story, and it can wait. I'm exhausted," I begged off.

CHAPTER 17

"Being Teddy ain't bad."

I woke up to the smell of coffee for the first time in three months. For most of that time, the time in Basic Training, I awoke to the sound of garbage can lids being banged, and somebody in a Smokey the Bear hat yelling something I never understood.

This morning though, the smell of coffee told me I was back in the normal world. I resisted the urge to roll over and grab another forty winks, because I'd be wasting a precious commodity I'd been denied, the pure enjoyment of civilian life. So, I rolled to the end of the bed, farted once just because I could, immediately feeling bad about the fart. It was one thing to fart in an Army bed with Army linen, it was more or less made for that. But these were flowery sheets, scented with some essence not meant to be mixed with a fart. I took my trousers from the chair and waved them around to move the fart away from the bed. I thought I did a good job until my mother came into the room with a steaming cup of coffee, and the sweet smile she reserved for me. That is, until she ran into the invisible vaporized gut-gas I had just spread all over the room. She hesitated for just a minute, the smile froze, and then she continued across the room, gave me the coffee and a kiss on the forehead.

"Sorry," I said.

Smiling, she waved it off, or maybe she was waving the smell away, and said, "It's good to have you home. Dress and come down for breakfast."

Between bites of home-made blueberry cream cheese blintzes and sips of freshly made coffee, I filled my father and mother in on my past couple of weeks. We spoke on the phone each week, but it was skimming the surface on the Army stuff. This morning, I had all the time in the world to answer their questions, feeling safe and secure in the familiar kitchen. The Pentagon assignment relieved them, but there was still a bit of unknown. Sure, there was a good chance I'd stay at the Pentagon, but there was also the chance I wouldn't. We lived close enough to Saratoga Raceway and Green Mountain Park to know there's no such thing as a sure thing.

"I just have to do a good job and not make any trouble," I said as a way to end the discussion and give them the confidence I had for my immediate future.

Mom remained pensive throughout the discussion, only saying, "Oh dear" after my attempt at reassuring them. A curious comment, but moms always worry.

I took my third cup of coffee out to the covered back porch to have my first cigarette of the day. Before the Army, I didn't smoke. Now, it was a regular thing. This was better, delaying the first cigarette until after breakfast. Better yet, if I could break the habit when I returned to Fort Myer. This assignment might be good for my health, if the Army didn't send me to Vietnam to have me killed.

It was nice sitting in the yard. The Albany paper, The Times Union, featured three stories on the front page about the Woodstock thing. There was talk of canceling it because of the rain. Not because it would make the audience uncomfortable, but because they were concerned the electric instruments and speakers would start a fire or electrocute someone.

My dad and I talked about normal things, like we used to before the Army. He filled me in on the happenings of my friends. Two of my high school classmates had been arrested at an anti-war rally. Might be good to avoid the old hangouts with this haircut.

"What do you have planned for the weekend?" he asked.

"Nothing, absolutely nothing."

As I finished saying that, the first drops of rain fell. I couldn't help thinking about the hippies at that Woodstock farm. They were probably all so doped up, they wouldn't even notice the rain.

"Not even Ted's?"

I smiled and said, "Oh yes, I can smell the fish fry at Ted's already." As my buddy Les would point out: Fish smells like pussy and pussy smells like fish, and us Southern boys love both.

For me, Ted's Fish Fry is the best meal in the world! Granted, my world is limited, but I can't imagine anything better. Ted starts with fresh cod filets, and cuts the twelve-inch filets into strips about an inch thick and one inch wide, making both ends tapered. Then they deep-fry the lightly breaded pieces of fish. The magic part is while the whole piece of fish is fried to a crispy golden brown, the tapered ends become even crispier. This delightfully cooked piece of fish is placed in a soft, warm, hot dog roll with the crispy tapered ends sticking out. A gentle dusting of kosher salt follows a ladle of mild chili sauce.

I've developed several peculiarities over the years to heighten my enjoyment of this repast. First, I break off the crispy ends of the fish hanging over the bun and eat them separately, without cocktail sauce. Second, while I usually put ketchup on the side of my fries and dunk, I allow the culinary craftsmen to ladle ketchup directly onto the fries. Third, I must eat the fries with a little two-prong wooden fork they provide. Finally, regardless of whether I have a coffee milkshake or chocolate milk, I always get a cup of ice and pour my drink over ice.

We went to Ted's at two. As I sat eating my second fish fry, looking at my parents doting on me, I realized I was doing the same thing Teddy was accused of doing—going home for the weekend to be spoiled by his parents. After a momentary pause, I continued enjoying my meal, consoling myself I didn't bring home a duffle bag full of dirty laundry.

Back at the house, Mom asked the same questions a second time. I suggested we wait and see what the future would bring. There was nothing they could do about it, and all I could do was work hard, stay out of trouble and prepare for a chance to stay in 1D726, which evoked another "Oh dear!" from Mom.

"I better take the train back to Washington tomorrow. Can't take the chance I'll get stuck in a traffic jam, and not get back. It wouldn't make a good impression."

The train ride back was uneventful. I avoided the traffic from the Woodstock event. In years to come, I could claim to have been there, having been stuck in traffic on the way up. I doubt if it will mean anything in the future. Who would want to say they were dumb enough to go an outdoor concert with a bunch of smelly hippies on a farm in the rain?

CHAPTER 18

"Learning More"

When I arrived at work on Monday morning, Henderson told me I had been reassigned. Before I could react with anything more than an increased heart rate, he relieved me of any further anxiety.

"Frank returned from leave early," he explained. Frank was his boss. "Sure wish you could stay and work for me, because I think you did a great job last week. Frank is set to get out in six weeks, so he'll be worthless as he gets shorter." "Shorter" was the term soldiers used to describe how close they were to the end of their enlistment. The shorter you got, the fewer days you had.

"Thanks. You're a good boss, and I enjoyed working for you."

"Mrs. Ramirez and her AIT assignment team needs help. One of her people is on leave for the next two weeks. The Major said she heard you were doing a good job for me, and put in a request. The assignment is only two weeks, and it'll expose you to other work done in the office. Good experience."

Andy introduced me to Mrs. Ramirez. After welcoming me, she introduced me to the rest of her team. They were all civilians, and she stood on formality. All were introduced as Mister or Mrs. or Miss, and I was introduced as Private West. The group assigned to her was large, about a third of the office civilian workforce. Offering me a seat next to her desk, she described what she wanted me to do.

Mrs. Ramirez was a large woman of about fifty. She wore a modest flowered dress, which I would learn was her uniform. She had them in various colors and emphasized different flowers. Her hair was short but beautifully styled. She wore no makeup except for a touch of lipstick and nail polish. When she spoke, her soft, well-metered voice had a soothing quality. Large glasses hung from a cord around her neck. When her glasses were on, her soulful eyes were magnified. Mrs. Ramirez was a Spanish version of my mother.

As she explained the responsibility of her team, she continued to refer to me as Private West. When I asked her to call me Jonathan, she took her glasses off and looked directly at me. In that same soft, yet powerful voice she said, "Private West, we have a very serious job on this team. We make assignments for all the young men who graduate from the Army AIT schools for which we have responsibility. We make these assignments against worldwide requirements, but most of the soldiers will go to Vietnam, into harm's way, and some will not return. So, we do our job with respect and complete professionalism. The ladies and gentlemen on my team are asked to dress in an appropriate manner. I ask each of us to treat the other with respect and in a professional manner. They will all address you as Private West until you are promoted. I ask that you address each of them as I introduced them to you. Is that clear?"

"Yes, ma'am," I responded.

"It's not necessary to call me 'ma'am.' Mrs. Ramirez is sufficient."

She explained that, at the beginning of each month, the computer across the hall spits out a new batch of requirements where soldiers are needed. These requirements are in the form of IBM punch cards, with eighty-eight characters coded to identify a job, the rank of the job, the location of the job, and any special requirements for fulfilling this job, such as security clearance. All the cards requiring Corporals/Specialists and Sergeants to fill the job go into a filing cabinet near the front of the office with a large SECRET sign on it. The rest of the cards are delivered to Mrs. Ramirez's team.

As for the AIT School graduates, each Monday the computer prints out a batch of IBM cards, one for each soldier who will graduate. All these cards are delivered to Mrs. Ramirez's team, where they are divided up according to school specialty. For example, someone handles the cooks, someone else the MPs, medics and another the clerks.

When she described the IBM cards for the jobs, she did it in a very matter-of-fact manner. These were just IBM cards. When she spoke of the IBM cards for the AIT graduates, she softened and spoke with a reverence. She never referred to these cards, or the people they represented, as assets or personnel or cooks or clerks. They were always "soldiers," or if she needed to be specific, "the soldiers from the cook school at Fort Dix," or the "soldiers from the medic school at Fort Sam Houston." When I slipped up, I got a look from above the glasses riding low on the bridge of her nose. The look was the same one the nuns gave me when they caught me smiling in church—wrong, inappropriate, and lack of reverence.

The job and her rules were easy enough to learn, and I got into the swing of things quickly, but Henderson was right when he said it was routine. As a potential lawyer, he should have a broader grasp of the English language. I would have chosen boring, mind-numbing, and monotonous as adjectives for routine. Any or all of them would have worked.

There were two distractions to help fight off the cobwebs building on the functioning neurons. The first was watching the rest of the office, the soldiers and the civilians. Mrs. Ramirez's team sat in the farthest corner of the big office, and unlike my desk with Andy, my desk in this group faced out. I could see ninety percent of the office and people, including our Branch chief, the Lieutenant Colonel, and the Major—the Assignment Officer. And then there was the Command Sergeant Major, or CSM, who was really the guy in charge. The remaining eight soldiers were like me—not there by choice, but grateful for what we had.

The Branch Chief was not a likeable guy; not that he did anything to make himself unlikeable, but he just seemed awkward and uncomfortable. He looked like he knew the CSM was in charge, and was struggling to figure out what his job was. During my first week, some of the guys referred to him as Colonel Klink, a reference to the awkward Commandant in "Hogan's Heroes," who also struggled with identity.

The Assignment Officer was a different case altogether. He had a friendly, outgoing manner, and was at ease with himself and everyone around him. The Major did everything the Colonel was supposed to do, and when he didn't know how, he'd ask the CSM. Age-wise, he was much younger than either the Colonel or the CSM, which put him less than a generation older than most of the draftees in the office. Claiming to be from Florida, he had an accent that sounded more like the hills of western Georgia. An easygoing guy, he got the respect of all of us, and no nickname.

Watching the other soldiers in the office interact, it was easy to see they were a group that worked as a team and played as a team, even though they had different jobs in the office. These young men had either just graduated from college, or had been grabbed out of graduate school or law school, like Andy. While they did their jobs as professionals, their recent extraction from a college campus was undeniable, and their interactions with each other and their manners were carefree and casual. The only thing missing, as they went about their work, was a can of beer in their hand.

The civilians were a different story—not Mrs. Ramirez's group, but the other half of the civilian staff. They were the reason soldiers had been added to the work force of this office, and the reason government employees have a bad name.

They had routines that would not be shaken, coming in at eight o'clock, and then going for coffee. Not just coffee: They went for breakfast, and then brought it back to their desks. During the next two hours, they'd do a few pieces of correspondence, and then go on break. Back from break, they'd do a few more pieces of correspondence, and then off to lunch. Same routine in the afternoon. They'd each put out eight pieces of correspondence a day. To put it in perspective, Henderson had been putting out about twenty pieces before lunch. Wednesday night was planned overtime, so the civilians could catch up if they had to, and they always had to. On Wednesday, they'd put out six pieces during the day, and save the other two for the three hours of overtime they worked after their paid dinner break.

The other distraction was Julie McFarland. Julie worked for Mrs. Ramirez. She was pleasant, but shy when introduced. She had short blond hair, dressed modestly, and following example rather than rule, she didn't appear to be wearing any makeup, except for lipstick and nail polish. Two things made her stand out. Julie wore Shalimar perfume, one of my favorites. Sitting only a few feet in front of me, she bathed me in the fragrance from day one. She didn't use a lot, but it was enough to recognize and get my attention. The other standout was her body. Regardless of how modestly she dressed and carried herself, there was no denying the fact she had a killer body.

The two weeks with Mrs. Ramirez's team became four. The person on vacation came back, but I stayed on. Henderson's boss, Frank—back from leave and counting the days—as predicted was worthless. He began a four-week process of saying goodbye. He came in late, claiming to have been invited to have breakfast with someone; took long lunch hours, claiming to have been invited out by someone; and left early, claiming a cocktail party was being arranged. Strangely, none of the people who worked with him in the office were ever invited to these parties. Frank was at his desk about three hours a day. During this period, I found myself sharing my lust, between a desire to have Julie sit on my face, and a desire to sit in Frank's deserted chair, which I now coveted as mine.

As I learned my job, I developed a conscience about what I was doing. I was safe in the Pentagon, more assured I wouldn't go to Vietnam. Yet, every day I was assigning young men, some with wives and children, to an uncertain future in Vietnam. I began to cheat. I stole assignments for Specialists in Europe, Panama, Korea, and Japan from the SECRET file, and sent the young soldiers with families to these safer assignments. This took a lot of extra time and work. The work part should be obvious: I had to decode eighty-eight columns of data from the IBM assignment card, and match it to an AIT soldier card, also with eighty-eight columns of coded information, all by hand.

Because I was falling behind, I decided to work a Saturday morning. I asked Mrs. Garcia if I needed permission. At first, she looked at me over the top of her glasses. Then she took them off and let them hang from her neck. She asked if the work was too difficult. I responded I was just trying to be careful with the assignments I was making for the soldiers who were graduating. The recent graduating classes had been larger, and the work load marginally increased. She kept track of this kind of stuff. She agreed I could work Saturday, and told me no special permission was needed.

CHAPTER 19

"Good Morning, Laos"

Working for Mrs. Ramirez was a hassle, me assigning guys like me to Vietnam. The Nixon administration's withdrawal of troops from Vietnam proved to be a fairy tale. If someone was taking them out, I was putting them back in just as fast. Phase 2 withdrawal had a target of 280,000 U.S. troops in Vietnam by the end of the year. Nixon and Laird told the newspapers and television the withdrawal was on target and effective, but no one had told the guy who kept punching out the IBM assignment cards.

Most people looked forward to autumn in Washington. The tourists are gone, Congress is in session, and the small-town flavor of this big city easier to enjoy. It was my second month in Washington, and I hoped to go into town to see some sights. But the past week had been heavy with assignments, and I had fallen behind.

I was sending hundreds of guys to Vietnam every week. These guys didn't want to go any more than I did. As each week passed, I saw the draft boards had finished with the bottom of the barrel, and were now scraping the sides and getting into the crevices. At first, the codes on the IBM cards listed most of the guys coming out of AIT as young and unmarried. I could save the married guys with families for Europe or Stateside assignments. For those I sent to Vietnam, I was buoyed by the fact the man-boys I sent were relatively safe. They weren't front-line soldiers, but clerks, cooks, lab technicians, and computer technicians who wouldn't see Charlie.

But as the draft boards picked up more married guys with families, so did I. It was difficult to send a married guy with kids to Vietnam, especially a draftee. It was easier if the code showed he had enlisted. At least he knew what he volunteered for.

So, I worked my first Saturday. This week's assets (guys getting out of AIT) versus this week's assignments (jobs in Vietnam) presented a problem, because we had more assignments for Vietnam than guys getting out of AIT. I wouldn't be able to protect the guys with families.

No rush this morning. Instead of breakfast at the mess hall, I grabbed a coffee, Danish, and a paper from the kiosk on the C Ring of the Pentagon, to take to the office. A simple luxury, being able to read the paper at my desk before starting work. A simple luxury, yes, but also a taste of reality, literally a taste of civilian coffee. Inside the Pentagon, all office coffee was Army coffee. Everyone got their coffee recipe from the military: one-part cat piss and one-part diluted crankcase oil. There's a sign on our office coffee pot: "Only stir twice or you'll be charged with destruction of government property—the spoon."

Skimming the front page of the Washington Post, my attention focused on an article about the previous day's press conference with Secretary of Defense Melvin Laird. A reporter reminded him of an earlier statement by the administration, which categorically denied the involvement of any U.S. troops in Laos. The press had their teeth into something, and asked him the same questions inside and outside, but really just calling him a liar—Melvin Liar!

Poor old Melvin. The freedom of the press seems to have gotten out of hand. Amazing! What happened to respect for the Office? No other country would allow a government official of his status to be called a liar. I felt sorry for him. He's probably making one-tenth of what he could make in the business world, and he has to put up with that crap.

"Tell them to fuck off, Melvin," I said out loud to all the people who weren't in the room with me. Another luxury of working on Saturday.

The jangling of the telephone jolted my thoughts. While a common occurrence Monday through Friday, on this Saturday morning it was an intrusion.

"General Support, West," I answered. I usually answered my phone with my "desk" designation, "General Support, EPCMR-GS, Private West." I never had to answer anyone else's phone; there was always coverage by another team member.

"Gee, General, I'm sorry. I thought this number was for Assignments Branch," said the voice through crackling and hissing of a long-distance call.

"No, sir. Relax, this office is GS, the assignment section for general support MOSs, not General Supports office," I replied.

"Ah, good, sir. I've been trying to get through all day. This is Sergeant Jenkins, and I'm scheduled to DEROS next month, and haven't received an assignment yet." My mind jumped to define DEROS. Ah yes, Date Expected to Return from Overseas assignment. The Sergeant was still talking. "Figured I'd check it out as long as I haven't heard yet, and I'd like to see if I can get back to Fort Bragg."

"Sergeant, with a DEROS of next month, you should be on orders already. Check with your clerk, and if he hasn't heard, have him check with your unit personnel and—"

"Shit, sir, that's a problem for me right now, 'cuz we don't have one with us. We're just all grunts. We been out here for a month now, and I can't get in touch with my unit clerk. I'm supposed to be here another two weeks, if I don't get killed, and by then it'll be too late to do anything about getting back to Fort. Bragg."

"Sergeant, if you can contact me, you can contact your unit, and they'll straighten you out."

"Sir, let me ask you. Are you enlisted or an officer?"

"Enlisted."

"E what?"

"E-2."

"Fuck me! I got me a piss-assed Private sittin' behind a desk in the Pentagon, telling me to scratch my ass while I'm laying it on the line here in Laos."

My eyes jumped back to the front page of the Washington Post. Melvin denied any troops in Laos, and I had a guy on the phone calling me from Laos, or at least telling me he was calling from Laos. Maybe it was one of the guys in the office pulling my chain. They knew I'd be working this morning.

"Is this a fucking joke? Is this you, Les? Just because I'm working today, you don't have to bust my ass. It was nice and quiet here until the phone rang."

"My name isn't Les, boy, and I don't give a shit whether you have to work. When they sent me into Laos, they said no contact with USARV or MACV or my unit. They didn't say nothing about not calling DA."

"Are you on the level? Are you really calling from Laos?"

"You bet your chair-borne ass, boy. I'll send you a postcard as soon as we overrun a souvenir stand."

"Jesus, you really are serious, aren't you? You're calling me from Laos."

"Look boy. If I knew I was going to this much trouble convincing you, I would have called you collect."

"Sergeant, where the hell are you calling from?"

"Laos, boy. I already told you that."

"No, I believe you, but where in Laos, and how did you get an Autovon line?"

"Shit boy, I don't know where in Laos. They didn't give us a Triple A trip-tic when we left. None of the road signs are in English, just some squiggly writing. What the fuck difference does it make?"

"Sergeant, stop calling me 'boy,'" hoping the irritation I had with him came through. "I'm looking at this morning newspaper, and Nixon and Laird are telling everyone we don't have any troops in Laos."

"If you transfer my call to one of them, I'll straighten the sons of bitches out, but that probably won't get me to Fort Bragg."

"What's your MOS, Sergeant?"

"Eleven Charlie forty."

"Okay, I believe you're in Laos, but I can't help you with your assignment. You're Combat Arms, and I'm in the General Support Branch. I'll transfer you to Combat Arms."

"Wait a minute, do I have to go through all this shit again?"

"No, just identify yourself and tell them where you're calling from. I'm sure they'll be anxious to help you," I said as I looked up the number for the direct line for the Secretary of Defense's Office in the internal Pentagon directory. He probably wasn't there, and if he was, wouldn't be answering his own phone, but it would be answered by someone in the know.

"Thanks, bo…, buddy."

"Hang on," I said as I depressed the button to signal the Autovon operator. "Operator, please transfer this incoming overseas call to 74972."

"Fuck him," I said to myself as I heard the click of the disconnect and replaced the receiver, thinking of three different individuals who qualified as the "him" in "fuck him." If it was a prank call, the prankster would shit when the SecDef's office answered. If it was Sergeant Jenkins and he was in Laos, he'd shit when SecDef answered. And, SecDef's office would shit when Sergeant Jenkins told them he was calling from Laos.

I returned to my newspaper, rereading the "no troops in Laos" article, this time with a different view of Melvin in the liar's den. Looks like the freedom of the press and the rights of the citizens to know the truth were on a collision course with the rights of the government to lie. I had a bitter taste in my mouth, and it wasn't all because of the coffee.

"Fuck 'em all," I said to the empty office, as I turned off the desk lamp and headed for the door. My extra duty day was over after thirty minutes of newspaper reading and a long-distance call from Laos.

Monday morning started with our regular ceremony of removing three days from our ETS calendar. We all had one of these in various forms, but they all conveyed the same message. Like everyone in the Army who didn't want to be in the Army, we knew exactly how many days we had to ETS, Expected Termination of Service. To make it even more bearable, we didn't even count the last day. If you were one hundred days from ETS, you claimed ninety-nine and a wake up, presuming you would be a civilian by the end of the day, so it shouldn't count. The draftee calendar had seven-hundred-thirty spaces for the two years stolen from our previous lives. Each day you got to fill in one square, except Mondays, when you got to mark three squares—one each for Saturday, Sunday, and Monday. I had my calendar in the elaborate shape of the Pentagon, with the squares evenly distributed around the five rings of the Pentagon, and the last day being the Mall Entrance, out the door.

From that high point, things went downhill fast. The Major was pissed I hadn't finished the work I was supposed to do over the weekend. I was pissed he was pissed. It meant working a few extra hours straightening things out.

When a soldier worked overtime, it was an invitation to the civilians in the office to work overtime. Civilian overtime meant a soldier had to work. It wasn't a supervisory thing, it was just the Pentagon required a soldier with a Secret security clearance to lock the safe at the end of the shift.

Knowing I was working late meant some civilians would set work aside during the day, and then about three-thirty go to their supervisor and say they had to catch up. I probably could have finished what I had to do in a couple of dedicated hours, but now I faced a longer night full of distractions, and quitting time dependent on when they felt they had enough.

About twenty minutes into this already miserable day, I got called to the front desk area where the Colonel, the Major, and the Sergeant Major shared their cubicle. The Colonel was talking to a short hair in civvies. Andy said he was a spook from the 902nd Military Intelligence unit, the group responsible for the internal security of the Pentagon.

"West, were you here on Saturday?" asked the Colonel.

"Yes, sir. In the morning," I answered. Shit, did I leave the safe unlocked when I was in such a hurry to leave? No, I didn't open the safe. I didn't even start work, just the newspaper and the telephone call.

I didn't have to take the thought process much further than that.

The spook took over. "Did you receive any unusual calls on Saturday, West?"

"Just some guy who said he was calling from Laos and wanted his DEROS assignment," I replied.

"Isn't that unusual enough to report?"

I didn't know I was supposed to report anything like that. The call could have been a wrong number. "It wasn't my MOS, sir. I transferred him to Combat Arms. I thought it might even be a prank, sir."

"You transferred him to the SecDef's office, not Combat Arms. Wouldn't you call that out of the ordinary?"

"The operator may have made a mistake on the transfer, sir," the response being almost a question, begging him to accept my answer before I had to dig this hole any deeper.

"It might be a prank, yet you transferred it to SecDef? You left within twenty minutes of transferring the call. You left the Saturday morning paper on your desk, the Washington Post with front-page stories of Secretary Laird's press conference about troops in Laos."

Shit! Mother fucker! Cock sucker! He's got me. I come in for a little extra work on my time, and now I'm gonna lose this primo assignment and get my ass shipped off to Vietnam, or maybe even Laos.

The spook gave the Colonel his full attention. "Sir," he said. The Colonel gave the spook his full attention. Man, this was serious, a Colonel standing at attention in front of a civilian. These spooks have power. Maybe he's like James Bond, licensed to kill—kill a career.

He continued. "Can I take West with me for a more private talk?"

"Yes, sir." The Colonel looked relieved as he said those two words. I wasn't going to be his problem; I'd be the spook's problem.

I looked around. No one looked my way. Deniability. When I failed to return, and they found my body washed up on the shores of the Potomac, everyone could say they didn't see a thing.

Turning, the spook said, "Let's go."

Visions of rubber hoses and electric wires shocking my testicles ran through my head. Which would be worse? Beat it out of me, or wait until my balls became testicle fritters? Yes, I wanted to scream. I knew I was transferring the call to Laird's phone. Yes, yes, I said fuck 'em all, including Nixon and Laird. No, no I don't want to die. No, no I don't want to go to Vietnam.

I wanted to say all that, but didn't—couldn't. The spook was wearing his good cop face. He was throwing me off balance, asking about me, like old friends who hadn't seen each other in a long time. Yeh, yeh, I'm happy here. I don't do pain very well. I'll tell you all you want to know. Please don't put electric wires on my balls.

That was what I thought, I don't know what I said.

The interrogation room turned out to be the snack bar on the second floor near the Concourse. My spook was a masochist and a sadist! He drank his coffee black and assumed I did, too. After some idle psychological bullshit chit-chat to throw me further off guard, he started in on the nitty-gritty.

"So West, someone called you on Saturday and said they were calling from Laos?"

"Right, sir." I figured I better call him "sir" because the Colonel called him "sir."

"Do you know for sure the call came from Laos?"

"No, but—"

"Isn't it possible it was a prank?"

"Sure but—"

"Then the only two things we can be sure of is you say someone called you and said they were calling from Laos. Neither of which you can verify."

"Sure but—"

"Doesn't it make you feel a little ridiculous to be so duped by a prank call?"

"I'm sorry. Did you say ridiculous?"

The spook fumbled with his coffee, then responded, "Well, yes. You can see how this has gotten out of hand."

My anger short-circuited my brain and I said, "I'm going to say I'm sorry again, but please recognize I don't mean it in the least. You say I should feel ridiculous. And I say, for what? For thinking it was a prank even as I spoke to the guy? Oh, no, you must mean I should feel ridiculous for not reporting it. In which case, I didn't create the situation that's out of hand. You, or the people who hide behind you, created it."

"West, calm down and listen to what you're saying."

"Mister," I started, but realized I didn't know his name, "you're acting like a dick, so that's your name." Rather than give me his name, he smiled at my rant. I continued, "You come into the office this morning and make me look like an asshole, or stupid. Either way, could get my ass sent to Vietnam. Question my credibility in front of the Branch Chief, his Deputy, and the NCOIC, and question whether I should have reported a wrong number. Now you tell me it might have been a prank, and I should forget it because someone duped me. You seem to forget I told you I thought it was a prank. So, if I did transfer it to SecDef accidently, then someone up there must be mighty interested in this prank, because they got duped, too—or it wasn't a prank, and I wasn't duped. Just in the wrong place at the wrong time. And you, sir, not me, is the one kicking up a bucket of shit about this call. So, if you have a bottom line on your report, let it read: You convinced West it was a prank call. Then you can go to SecDef and convince whoever it was up there who started this cascade of bullshit, it was a prank. Mister whateveryourname is. This situation is of your doing, not mine. Remember, I did nothing."

"Allen."

"Allen?"

"My name is Allen, Mike Allen, not dick," he said calmly. Calm enough to make me think I might be behaving in a way where the Psych 101 handbook says he has to be calm if he wants to talk me off the ledge. "Calm down, drink your coffee," he added as he slid my coffee closer.

"You slide the worst-tasting coffee in the world to me to calm me down?"

He shrugged and took a long pull on his coffee. Remarkable, his face didn't squint at all. I couldn't say the same for mine as I watched him drink.

"Mister Allen, this stinks. The coffee stinks, too. I'm outta here, before I say something I'll regret when I'm sleeping in the mud in Vietnam next month. I've got work to do. So do you."

I left the snack bar and started back to the office. Fuck the 902nd, fuck the spook, fuck Melvin, fuck Nixon, fuck Sergeant Jenkins, and fuck the Army.

What a terrible feeling. It was that pivotal moment in my life when I stopped believing in the government, when I questioned everything I read in the paper or heard on the television. The collision took place in my head that day, but got played out on the Formica table top between me and Mister Allen as he attempted to have me redefine the truth.

I should have nodded agreement with anything he wanted me to do. I hope I didn't screw up this cushy assignment.

The last laugh is always the best.

A few weeks later, the Washington Post announced elements of the U.S. Army had been deployed in Laos and Cambodia to break up staging areas and supply routes the North Vietnamese used to wage the war in South Vietnam.

I wonder what happened to Sergeant Jenkins, and if he ever got to Fort Bragg.

CHAPTER 20

"Learning the Magic"

It was well into October now. The weather was changing in Washington, as was I. They built Washington on a swamp. With the swamp and the river, the humidity in the summer became unbearable. The city was tolerable only because of air-conditioning. When chosen as the seat of our government, there wasn't any air-conditioning. A cruel joke, or good planning by our early Congresses who wanted the summers off? They have described the District's weather as having an average humidity of fifty percent. When challenged during the summer, the glib answer became, "zero percent in the winter, one-hundred percent in the summer, on average fifty percent." In October, the normal was achieved.

But, while the weather became more pleasant and less oppressive, my job became more oppressive. Sending young men to Vietnam became more troubling with each soldier I sent. I was still waiting for my Top Secret security clearance. The incident with the Laos telephone call shattered my belief in the system, and may have reduced my chances of getting the clearance.

While still confined to the formal work behavior dictated by Mrs. Ramirez, I could see the rest of the soldiers in the office who didn't work with those restraints. They took their work seriously, did it well, but had fun doing their jobs. Every job took into consideration the soldier represented by the IBM card or the file they held. Everyone else was bringing them home or fulfilling requests soldiers had made. But I was the only soldier in the office sending guys to Vietnam. The more I thought about it, the less I liked it. Working with Henderson had been better, because he brought a lot of guys home. I decided to have a talk with him.

Returning from lunch, I asked if he had a few minutes.

"What's up?"

I explained everything, with brevity and clarity, focusing on my concerns of sending men to Vietnam and the disillusioned view developing in my mind about the government.

Andy spoke calmly, "I guess I had a leg up on you coming into the Army. Science really is an ivory tower, and there's usually only one answer. History and law are different. We study the mistakes made in the past. The big mistakes are made by the government. Being in school in New York also exposed me to more radicals and protesters than you saw. I had no illusions about our government or the war. The war isn't a just war, and our government is lying to us daily. I can't change that. I can only do my job, move on, and hope I'm prepared to help make changes in the future. Until then, I help the guys I can."

"What about my job, and waiting for the security clearance?"

"That's where I can help you. I can spin the security clearance so it's good for you, good for me, and good for the office."

"Really? How's that work?"

"I get out next July. The Major asked me to think about my

replacement. There's no one in the office with enough time left to train, so that means I have to get someone from the outside, probably Fort Dix clerk school. You have a lot of time left, you're perfect."

"But what about my security clearance?"

"So what? Secret is all you need here. All the Army guys, including the Colonel and the Major, will be gone when you get out. I'll sell you as the continuity bridge to the next officers to run the office."

"What about the Sergeant Major?"

"Oh, he might still be here. Word is he'll retire out of this assignment. He's over fifty-five. The Army's already extended his retirement age once, and they won't do it again."

"Will they want me?"

"They've already said they hope you don't get the clearance."

"Really!"

"Not in so many words. More like 'too bad West can't stay.' It'll be okay all around. Even for me. I've already got you half trained."

Andy surprised me. Five minutes ago, I felt low and confused, struggling with what was right and what was right for me. Now, all seemed right with the world.

"Let me talk to the Major and we might wrap this up today. Who knows, I might even let you write your own orders. Then you can get an apartment and start living the good life. Well, that might be an overstatement, but you get the picture."

Within an hour, I was back at my desk by Henderson. To his word, I typed my orders transferring me from TDY to permanent status in the office. Ten minutes later, the Colonel signed them, effective immediately. The Major called the office together and announced my assignment. Stoner gave me a half-smile and a half-salute.

The next day, I settled into the familiar routine of working with Henderson. He'd review the paperwork on his desk, scribble on a note card, and send it across the non-space between us for typing. If it was out of the ordinary, he took the time to explain it.

At my desk, I couldn't see the office as well as when I worked with Mrs. Ramirez. I missed watching the fraternal dynamic and the back of Julie, but made up for those losses by being more in the center of a dynamic. I wasn't part of it yet, but it operated all around me. Before, I could only watch it. Now, I could hear it and feel it.

I hadn't noticed it before, but Henderson was the core of the office. Everything important revolved around him. Sure, he had the job he was supposed to do. That was the part I had seen before, and all I paid attention to. Now, as his heir apparent, I paid attention to everything. There was the one world of the office, but it had two different orbits. The first orbit was the fraternal dynamic I had observed from afar. It only included the enlisted men, excluding Stoner, and Henderson was the core of the fraternal orbit. One could argue it was a small group, but that argument, while true, didn't reflect his influence. He was dynamic in everything he did, even to the bounce in his step. Everyone sought his acceptance. Those he accepted into his sphere were few—Mullaney, and now me.

In the military sense, he was an unlikely leader. Short, bald, and a little overweight, he had an almost perfectly round face. Peter, an impish young man from the Upper Midwest, had tried to give him the nickname "Ziggy," after the cartoon character that had appeared the previous year in greeting cards. Henderson said no, pointing out Ziggy didn't have a moustache. Peter persisted for two weeks, but then dropped it.

The second orbit was the office work, which included all the office activities, and the half-million plus men we handled around the world. This orbit was larger and more inclusive, involving all the people in the office: civilians, military, enlisted, and officers. Here, Henderson was also the core. Everyone valued his input. Beside the work I was familiar with, Henderson got all the special assignments that required thought, research, diplomacy, or all three. Now I noticed regular calls from the Colonel and the Major for Andy to go to their desks. The Sergeant Major wasn't someone who shouted from across the office. He always came to Henderson for his input. Some of this extra attention resulted in extra work. Sometimes, all they needed was his opinion. I had doubts of being able to fill his shoes.

"Andy," I said, "I've been paying more attention this week to what you do, and how you do it. Seems you're involved in everything, but I'm only familiar with a small part of what you do. I'm sure I can handle that, but this other stuff seems more complicated."

"Don't worry about it. You'll catch on."

"Yeah, I know I can do the stuff you push my way, but I'm concerned about all the special stuff the Colonel, the Major, and the Sergeant Major ask you to do. And then there's the stuff that comes walking in the door."

"Like what?"

"Like White House Liaison, Congressional Liaison, and the stuff you do for the spooks in the 902nd."

"You'll get it. It's nothing special. Some stuff they ask me to do, they already have an opinion. They just want a second opinion. They think of me as their in-house lawyer."

"But, I'm not a lawyer."

"Good, you have nothing to worry about. They won't ask you."

The timing was perfect to illustrate my concern. A Major from the White House Liaison Office down the hall approached Henderson's desk. White House Liaison had two offices: one in the White House, and one down the hall. Their job is to respond to the people who write letters to the president about soldiers. There are lots of mothers out there who are not happy with the way Nixon is handling the war, or maybe it was just the way he was handling their baby boy. They write a letter to Nixon, and he doesn't know the answer, so the White House sends it through the liaison office, and it winds up for Henderson to handle.

"Henderson," said the Major as a way of greeting.

"Major."

"I've got one that's a little unusual for you, Andy."

"You always do, sir. Do you mind if West listens in on this? He'll be flying this desk when I ETS."

The Major agreed and described the special case.

"We have a young Specialist returning from Vietnam who's been assigned to Fort Hood and would like to be with his family in Florida. He didn't write the letter, his father did. The father is president of one of the major television networks and wanted to use his pull to get the reassignment. The young man in question put in for a Permissive

Reassignment when he first got his DEROS orders, but the response from DA hasn't caught up with him yet."

Andy explained for me, "The Permissive Reassignment is one of two kinds of action a soldier could request, the other is a Compassionate Reassignment. A soldier can request a Compassionate Reassignment if he has a compelling circumstance that he feels the Army should take into consideration. These usually involve a medical or economic hardship which his presence would help ease. For example, a sick parent or child who needs special treatment. If the Army agrees, they would move him at the Army expense. A Permissive Reassignment is slightly different. It could include a personal hardship, but could also be because the soldier didn't like his assignment and had someplace else in mind. The big difference between the two is, with a Permissive Reassignment, the soldier pays all the costs of the relocation out of his own pocket."

Andy turned back to the Major and asked, "What's his MOS and service number?"

The Major flipped through the file. It was extensive because all documents had to ascend the Chain of Command, with each step along the way adding paper. The Major found the information and provided it to Andy.

"Sir, I believe he's already been assigned to MacDill Air Force Base. The Army has a small support function there that's often overlooked when it comes to staffing." Andy went to the SECRET cabinet, searched a little, and produced an IBM card. He took out a pen and made a notation on the card and gave it to the Major.

"He was in the batch processed yesterday. The approved Permissive Reassignment paperwork, and reassignment order, should arrive at the same time."

As he left the office, the White Liaison Major said to the Colonel, "Sir, never let Andy go. I don't know what I'd do without him."

When Andy came back to our side of the office, I stopped him before he could sit down. "I can't do that."

"Can't do what?"

"Memorize the service numbers of all the soldiers we have responsibility for."

"Why would you want to do that?

"Because, you just did it?"

"No, I didn't. Asking for the service number is just how I stall for time to think. Sometimes it gives me the chance to peek at the papers the Major's carrying. Maybe see something that jogs my memory."

"I don't get it. It looks like a magic trick."

"No, not magic, but it's simple. It's the logic we use in the law— sometimes. The scientist in you will like it. First, we're not talking about everyone in the Army, we're only talking about the small group we're responsible for."

"That's still a lot. Just the guys in 'Nam are about 500,000."

"See, there you're wrong. That 500,000 represents all the U.S. servicemen in Vietnam: officers, enlisted men, Army, Navy, Marines, and Air Force. Let's say, for the sake of argument, 200,000 are Army, and of the 200,000, 25,000 are officers. That would leave 175,000 enlisted. Let's say another 25,000 are E-7 and above. That leaves 150,000, and I think that's a high number. The three major MOS categories are

Combat Arms, Maintenance and Transportation, and General Support. General Support is a smaller percentage, say thirty percent, so we have responsibility for roughly 40,000 soldiers in 'Nam. Are you with me?"

I answered in the affirmative, not knowing where this was going. I also didn't know if his assumptions were correct, so I was only agreeing to the arithmetic.

"These 40,000 soldiers are there for twelve months, so that means every month there are about three thousand soldiers who go over to replace the same number coming home."

I could do the math in my head, and he wasn't answering my question. "So, instead of 500,000 service numbers, all I have to do is memorize three thousand this month, and then a different three thousand next month?"

Andy ignored the sarcasm in my question. "No. Most of those guys coming home accept their assignments, and we don't hear from them or about them. We seldom hear from the guys going to 'Nam. If they have a problem, they go to the Chaplain or the Red Cross, and they take care of it."

"So, how many guys are we dealing with—you know, guys who come back and have a problem with their assignment?"

"Not that many. Probably about one hundred guys come back every month without an assignment, or have an assignment to someplace they don't like. Some will only be there for six to eight months, and they figure they can do that time standing on their heads after being in 'Nam. The guys who come back with over a year are the ones who enlisted, and they're a little pickier. We try to accommodate them, and you'll get a chance to see how we do that."

"What about the Major's White House guy? How did you know about him?"

"When a guy in Vietnam gets his orders to return and doesn't like the assignment, he goes through channels. It winds up on the desk of an HQ personnel guy in Vietnam. They call every morning at eight, and every afternoon at four. 'So and so is DEROSing in three weeks, and they've assigned him to Fort Lewis, Washington, but he wants to go to New York City.' Every morning during the call, I make a list. Then I check for assignment cards in the file, and if there's a slot, he gets it. If not, sorry. Sometimes the personnel guy in Vietnam has talked to the guy and knows the background story; a wife, kids, mother, job, or school, which makes it special. If that's the case, I make up an assignment at the place he wants to be, and send him."

"You make up an assignment?"

"You'll learn how to do it. The only ones in this office who can do it are me, the Colonel, and the Major, and now you. The Colonel and the Major don't like to do it, so it's just us."

"So, what about the White House guy?"

"If I have nothing for the personnel guys in Vietnam, they tell the guy to submit a Permissive Reassignment request. We talked about that already. He does it right there and avoids part of the chain of command. The paperwork gets on my desk about the time he gets back in the States. After he leaves Vietnam, he checks in with the Personnel guys at the entry point, like Oakland. They check on the status with me. If an opening jumped up, we cut him new orders. If not, he takes his thirty days leave and checks again when he shows up at his new post. After thirty days, if it's still a no-go, we rarely hear from him again. This White House guy just showed up back in the States."

"So, how did we learn about the White House guy?"

"Most often, it's the mother, father, or wife of the soldier who writes the letter to the president, or to the member of Congress. Congress has their own liaison desk, called Congressional Liaison. They think there will be more weight on it if it comes from a big wig; but the bottom line is, all these things wind up here on my desk. If only they knew."

"So, what was the magic with this White House guy? How, or why, did you remember him?"

"No magic. I saw his name three times in the last five weeks. If I can't remember something like that, I don't belong in law school, or at this desk. I got the request from Vietnam before he left, and a follow-up from Oakland when he showed up there, and I reviewed his Permissive Reassignment request. I remembered him because he listed his hometown as Newport, Rhode Island, a pretty ritzy address. He enlisted after two years at Yale, and he requested an assignment in the Miami area. That sticks out in your mind."

"You read all of that information in the paperwork?"

"Yes, it's the details that are interesting in this job. The work is easy; we have an assignment, or we don't. The background in the guy's file fills in the details, but still leaves questions. I remember thinking, 'Why would a guy who seems to have the world by the ass—rich kid from Newport, having a good time at Yale—want to enlist in the Army, when he knows he'll go to the 'Nam?' And then, to top it off, wants an assignment in Miami? Makes you wonder, doesn't it?"

"And…?"

"So, when the Major came in, he filled in the answer. The father is a big wig. The kid was a rebel of some sort. Flipped off the college bit.

Flipped off his parents, and went off to war. Miami is probably where his family has their winter home. He'll have two winters before he gets out and has a chance to spend more time with the family he flipped off. Plus, he'll get a nice tan before he goes back to Yale."

"The guy was lucky you found him a slot at MacDill," I said.

"Nothing lucky about it. I made up the OPO Control number, and made the slot for him. The kid earned it. He took a chance. Glad I id. Looks like his dad forgave him and tried to do something good for him. And, as they say in the funny papers, 'everybody lived happily ever after.'"

"And the Major thinks you're a genius."

"The legend lives and grows, to be passed on to you," he said smiling. "Are you up to it?"

"Sure am. Bring on those angry fathers, mothers, and wives who think the Army is screwing around with their family. I can't wait."

I wouldn't have long to wait before they would test me.

CHAPTER 21

"'Oh Dear,' Defined"

When Les, Kennedy, Dexter W., and I showed up at 1D726, we were just four guys passing through. While they moved on, I waited for a security clearance. Even though I had a chair and a desk in the office, I was still a transient, and soon gone. The staff, civilian and military, tried to make me feel welcome, asking about my background, where I was from. Sure, they were polite, but indifferent. This changed after I was permanently assigned to the office. People who had politely inquired about my background asked the same questions again. Now, they expressed an interest in my answers, the same answers provided before. The difference? Now, I was one of them.

Sitting in my temporary chair, at my temporary desk, doing my temporary job, was like being at a baseball game, in the first row behind the home team dugout, with my glove and home team hat. I was close to the action and enjoyed watching it. But now, they put me on the team. The glove and hat became tools, not decorative adornments.

Besides me, someone else worked temporarily awaiting a security clearance. He was slated to work the courier service, transporting sensitive documents around the country or around the world. Both being transients, we should have bonded, but we didn't. Now, as I looked at him, I viewed him as they had seen me—a transient. No sense in getting to know him, he'd be gone soon.

I had become familiar with some of the guys. Andy and Mullaney had included me at lunch, but I was merely tagging along. The change to permanent assignment promoted me to an equal, and worth learning, about because I would be around for a while.

As I sat enjoying my new-found stability, Andy answered his phone. He paused, listening to the person on the other end, and then said, "Hello. Hello. Can you hear me?" After pausing a minute, he hung up. Seconds later the phone rang again, and he went through the same routine. After a third time, I asked if he was having trouble with his phone.

"No. It's a screamer."

"A screamer?" I asked.

"Andy explained. "Every once in a while we get someone calling in who's upset with something: an assignment, the lack of an assignment, or even no response. They scream into the phone. When they scream, they don't listen, so I just make out like I can't hear them. They keep calling back. When they're calmed down, I can hear them."

"How many calls does it take?"

"Four or five."

The phone rang again, and apparently this time the voice on the other end wasn't screaming. Andy began his conversation and then abruptly ended it saying, "I can't hear you again. There must be something wrong with this connection because every time you raise your voice, the line cuts out," and he hung up. Looking over the top of his glasses, smiling he added to me, "The next time he'll be fine."

Sure enough, the phone rang again, Andy listened and then transferred the caller to the Combat Arms group. Hanging up, he looked at me and said, "Wrong number."

Ray, the guy who was training to take over for Teddy, had been a History major, doing graduate work at the University of West Virginia when he got drafted. While I hadn't spent any time with him over the first few months, he invited me now to join him for lunch. Ray was a nice enough guy, about six feet tall, and putting back the weight the Army had taken from him during training. The front of his summer khaki uniform shirt was challenging the strength of the buttons around the waist. Fortunately, the Pentagon had no uniform of the day policy, allowing him to neutralize this challenge by wearing Class A greens all the time, even in the summer. He also challenged the Army haircut regulations with long blond hair constantly in his face.

"Are you going to stay in the Consolidated Barracks, or get an apartment?" he asked.

"I'm tired of living with a bunch of guys," I replied, recalling college, my time in minor league hockey, and the last four months in the Army. "In all that time, I've had to share living accommodations. Might be nice to be on my own for a while, or maybe one roommate. I've called about a few apartments in the paper, but when they ask about my income, my Army pay doesn't qualify."

"Might be able to help you there. I live in an apartment complex in Arlington, built during the boom in the late 1940s. Rumor has it, Nixon lived there when he first came to Congress. The complex is about half military with families, and half civilians."

"Sounds good, but what about the rent?"

"My wife and I have a furnished two-bedroom. The rent's $121.00 a month, including heat and electricity."

"How will I get past the income requirement?"

"Just do what we all do, lie. Tell them you have 'other income' when asked. Tell them a thousand a month. The agent will smile, say okay, and hand you the lease. If you want, I can pick you up on Saturday, and you can take a look."

I accepted his offer, saw the apartment, falsely filled out the form, and joined the government conspiracy to lie if it was in my best interest to do so.

The Sergeant Major approached me after the weekend.

"Ray says we're going to be neighbors."

"I beg your pardon, Sergeant Major?"

"Ray told me you took an apartment in our neighborhood."

"Yes, I did, but I didn't know you lived there."

"Yes. Mrs. Stoner and I live at the bottom of the hill, two buildings over from him. We ride to work together. Before you ask, Mrs. Stoner is my wife. She's Australian, sounds snooty when she talks, like a Brit. I think she should be referred to that way, and she agrees. Our little joke. Do you have a car?"

"I assumed I'd take public transportation."

"Bad choice. There isn't any that's convenient. You'd have to change buses at least twice and the commute adds three hours to the work day. You'll ride with us," and then he was off. I watched with a dropped jaw.

Mullaney came through and saw me staring at him retreating. "Don't let him catch you watching him walk away. The old guy's been in the Army a long time, and he might have bad memories of guys watching his butt."

I snapped my head and eyes away and said, "I got an apartment this weekend near Ray. The Sergeant Major just told me he lives in the same neighborhood, and I'm to ride to work with him."

"Sounds like job security to me."

"But I thought just being in this office was job security?"

"True, but now you have more."

I was feeling pretty good. I had landed on my feet in the Pentagon, and would be here for the rest of my time, and scored an apartment I probably couldn't afford. It was just going to be a bad-paying job, less than three hundred a month for a forty-hour week, but no nights or weekends, and nobody shooting at me. And, I liked the job. Learning from Andy was broadening me, taking me out of the ivory tower of science. I'd be a better person for it. Confidence and contentment had replaced uncertainty.

The next morning, when the Major shouted out, Andy looked across the space between our desks and said, "Go see what the Major wants."

"Am I ready? There's one of the guys from White House Liaison up there."

"Only one way to find what you know and what you don't. If you know, great. If you don't, we'll see how well you fake it."

Reluctantly, I went over to the Major's desk, where he introduced me to Mr. Durnan, a Warrant Officer from White House Liaison. Warrant Officers were sort of officers, and sort of enlisted men, occupying a place between the two. I never understood why they were needed or how they came to be. They were best known as helicopter pilots in Vietnam. The Army might have figured they needed someone other than an enlisted man flying an expensive piece of machinery, but didn't want to invest the money for an officer in a job having a life expectancy of twenty minutes in combat.

"Stepping in for Andy?" Durnan said with a tone that let me know he was less than pleased dealing with the second string. "I'm looking for a guy named West." He chuckled. "Must be a coincidence I get to work with you."

Durnan was about my height and age. There the similarity stopped. He was slight, about one hundred thirty pounds, with blond hair that didn't match his brown eyes. He had a nervous energy, fidgeting with the papers in his hands, eyes moving around the office. Taking me aside, near the SECRET filing cabinet, he said he usually worked with Andy. I explained I had been working in the office, waiting for an assignment, but had just moved over to train as Andy's replacement.

"Those are big shoes to fill. Andy's a legend in this hallway. Don't know how he does it, memorizing all the service numbers. Must be a natural ability. Probably takes a lot of memorizing to be a lawyer, too. He'll do well."

I agreed, and he started talking about the case he needed help on.

"It seems we have a soldier who, according to his mother, may have been screwed over by his recruiter. He came into the Army from Albany, New York, and went to Fort Dix for Basic and AIT."

My heart rate jumped. West, Albany, with Basic and AIT at Dix? Coincidence?

"Then he seems to have dropped out of the system. Can't find him on any of the records."

As my heart rate continued to climb, I pulled the "Andy" and asked for his full name and service number.

"Let me see," he said as he leafed through the attached pages. Behind the formal typed documents from the White House to the liaison office, was the initiating letter from the mother. As Durnan searched for the service number, I had time to see the letter, albeit upside down. Written on stenography pad paper with pencil, the punched holes at the top of each page showed the page had been torn from the note book. "Here it is; the service number is 11965099, and the name is—"

"Jonathan West," I finished for him, as I understood why Mom had had uttered "Oh Dear!" when I said I'd stay at the Pentagon, unless something got screwed up.

I'd describe the look on his face in detail, but it would take far too many words; awe will have to suffice, as he said, "That's it."

If I had to describe the look on my face, "scared shitless" would do it. "I don't know how Andy does it either. I just know this guy's case," I said. "Can we go into the hallway to talk?"

"Sure, but you have to tell me how you guys do this. This is unbelievable how two guys with this ability wind up in the same office, doing the same job."

When we got into the hallway, I stammered, "I know that guy."

Durnan stopped speaking and stared. If awe described him before, now he was well beyond awe. "That's something Andy can't do. How do you know him? Do you know where he is?"

"It's me."

"Yeah, I know you're you. The Major told me. Now help me find him."

"No, I mean the guy you're looking for, West from Albany. That's me."

"This is un-fucking-real," he said. His hands had stopped fidgeting. His jaw dropped while his head came up. He did a little spin, not entirely on one foot, but mostly.

"Yeah, but look, I'm happy with my new job. Can't we just throw that letter away? Forget she ever wrote the letter? I didn't even know she wrote it."

"Is this story going to take a while to tell?" Durnan asked.

"The quick response is, when I got drafted, I went to a recruiter in Albany to see if I had any options. He took my background and told me if I enlisted, I could go into Special Services at Fort MacArthur in California, and play on the Army hockey team. When I got to Fort Dix, they told me the Army doesn't have a hockey team."

Durnan was sympathetic and said, "Sounds like a fraudulent enlistment. Let me work on this for a couple of days and I'll get back to you. I can't just let this go away. I have to answer the letter to the President. But, to put your mind at ease, I don't think any bad will come of it, and you might be pleased your mother took the time to write a letter to President Nixon."

"What about the office? The Major will want to know how I did.

This is my first test as Andy's replacement."

"That, my friend, I can take care of right now," and marched back into the office.

"Mr. Durnan, how did it go?" asked the Major.

"Well, sir. Andy's good, but if this first test is any sign of what West can do, he might be even better than Andy." With that he left, saying he'd be in touch.

"Well done, West. I guess that's why I'm such a good Assignment Officer. I picked you." He said it with a smile I took to be an acceptance.

As I left and returned to my desk, I heard him tell the Colonel and the Sergeant Major, "He done good."

Seated back at my desk, facing Andy, he asked, "How did it go?"

I slumped in my chair. "Well, I didn't have to fake it."

The fall of 1969 witnessed major events with impacts on the war in Vietnam and activities at the Pentagon. Ho Chi Minh died on September 2, and rather than a lull in hostility, his will, read by his successor Le Duan, intensified the conflict. Ho Chi Minh's last words urged the people of North Vietnam to fight until the last American was driven from the land.

The only time the Sergeant Major became reflective was in September, when the My Lai Massacre story broke, and the Army charged Lieutenant William Calley with the murder of civilian Vietnamese women and children in March 1968. Lieutenant William Calley was on trial. Journalists, both print and television, were screaming about him.

The trial gave the press another weapon to use in their war against the War and the Nixon administration. The atrocity, just another reason for us to get out.

Stoner had been in Vietnam immediately before being assigned to the Pentagon. He was already a Command Sergeant Major at the time. As one of the most senior enlisted men, and one of the most trusted, he was assigned to one of the numerous Army teams investigating My Lai. A young Major named Colin Powell led the team. Major Powell issued a neutral final report. Stoner had been a dissenter to that report, and his dissention had, in part, led to a more in-depth investigation by General Peers. The Peers report placed the blame where it belonged.

Stoner was a man of integrity.

CHAPTER 22

"Dear President Nixon"

Andy would be getting involved in this thing with my mother's letter, so I filled him in. Durnan had expected to be dealing with him when he came looking for help, not to mention he was my boss.

"Figured you could do the job, that's why I sent you up there. If I had done it, I would have said, 'I know where this guy is,' and pointed to you. Wouldn't have been as dramatic as you finding you. Let's see what happens."

"What will happen?"

"Don't know, but it won't be bad. The Major and the Colonel are okay guys, especially the Major. They'll figure it out and give you some choices. The final decision will be yours, and they'll stand behind you. Does that help?"

"Yeah."

"Thanks for telling me."

As we finished talking, the Major called for Andy. He went over to the Major's desk with the same easy manner as always. The Major asked him something. When he answered, the Major laughed and looked over toward me. The Colonel said something that looked like

"What's so funny?" The Major slid his roller chair over closer to the Colonel's desk. He looked like he was telling him a joke. The Colonel acted surprised then laughed and then said something to Andy, who also laughed. They called the Sergeant Major over. The Major appeared to be telling the Sergeant Major the same joke, because the smile was the same, as were the head and arm motions. Sergeant Major just raised his eyebrows, looked at me, and said something to Andy. The Major, the Colonel, and Andy all shared another joke, laughing and shaking their heads. Andy came back to his desk, grabbed a file, and started working.

"Well?" I asked.

"Well, what?"

"Was all that about me?"

"Yeah."

"Well?"

Andy sighed, put down the file, smiled and said, "They asked how you did."

"And?"

"I told them you found the guy, but some things had to be checked out. Then they asked me if you needed any help, and I said I didn't think so. The Major has been working with me long enough to know I was holding something back, so he asked what it was. I told him the guy Durnan was looking for was you, because your mother wrote a letter to Mr. Nixon. The Major told the Colonel and then we had to tell the Sergeant Major."

"That's it?"

"Not exactly."

"What then?"

"The Major said I must have done a real good job of training you. The Sergeant Major disagreed and said anybody could find themselves. All they had to do is reach behind them, put both hands on their ass, and there they were. The trick was finding someone else. They laughed and told me to get back to work training you."

"That was it?"

"Yup."

"What now?"

"I get back to work, and you wait for Mr. Durnan to get back to you."

"That's it?"

"Yup."

Within the hour, most of the Army guys in the office knew about what happened. No one cared, because no one, including me, knew what was involved in a fraudulent enlistment. No one asked me about it, except for Ray. Ray had been happy when I joined the office team because I became the longest, the one with the most time 'til ETS. There's an unwritten hierarchy in the military besides rank. Within the ranks of the lifers, everyone knows who the senior officer is. If there are twenty Captains, they know who's been in rank the longest, and he becomes the senior officer. The same thing holds for lifer NCOs. With non-lifers, the most admired are the shortest, fewest days left in the Army, while the goat is the guy who has the longest time remaining. With my arrival in the office, I had taken the goat distinction from Ray.

"So, I hear you might have a fraudulent enlistment action," Ray began.

"That's what they say, but I don't know what it means."

Ray was a history major, which in his mind automatically made him an authority on everything. Historians are like that, by nature. Everything had a basis in history; either it had happened in the past, or was happening now and would be history tomorrow. So, history was the seat of all knowledge.

"Looks like you're saying the Army misrepresented itself. It's like you're taking them to court."

I wasn't taking the Army to court. I just found out about this twenty minutes ago. All I was doing was waiting for options from Mr. Durnan.

Before Ray could exercise his historical histrionics on me any further, Andy, the lawyer-to-be, said, "Ray, back down. You don't know what you're talking about. White House Liaison is checking this out, and they'll offer him options. All will be good for West, and he'll choose the best. The only thing for sure is when this is over, you'll be the longest again."

Ray walked back to his desk.

"Ignore everybody until Durnan comes back with your options."

That's exactly what I did, but it was difficult. The next time I walked past the Major's desk, he stopped me.

As I stood in front of him, he leaned forward, his friendly nature not masked. "West, you sure made things exciting today. I can't wait to see how this turns out. It'll be all over the building by the end of the week."

"Thank you, sir."

The Sergeant Major, while a crusty old lifer, still had his soft moments. Aware of my anxiety about the future, he reminded me I would still ride with him to work when I moved into my apartment. I took that as a good sign I wouldn't go to Vietnam.

It didn't take long for Mr. Durnan to get back to me. He suggested he do the debriefing with Andy, the Major, the Colonel, and the Sergeant Major present, rather than having me tell them.

"I've run this up the flagpole, and everyone saluted," he said. Reacting to my puzzled look he said, "That means, I got agreement from everyone in my shop as to a way forward, if it's okay with everyone here. The first thing we have to do is respond to the letter West's mother wrote to President Nixon. It's an open file and must be closed."

"It's okay. I spoke with my mother on Monday night, told her about Mr. Durnan, and she's cool, now that I'm here. She said to tell you she's sorry if she caused any problems."

Durnan responded, "Sorry, it doesn't work that way. We've got to close out the file, same as anybody else."

Andy spoke. "I'll put a routine letter together, like we normally do, and get it into the system."

"No, in this case, I'd like West to draft the letter for my signature," said the Major.

"Good idea, but I'll sign this one, and we'll send the paperwork back up the chain to the White House," said the Colonel.

Andy smiled, the Colonel smiled, the Major smiled, and the Sergeant Major smiled. Everyone was smiling except me.

Andy and I went back to our desks. He told me to draft the letter and give it to the Major to review. When I asked if he wanted to see it, he said no.

"So, I'm going to draft a letter to my mother saying everything is okay, something she already knows."

"No, you're going to draft a letter the Colonel will sign and send to his boss, a one-star General, who will send it to his boss, a two-star General, who will send it to his boss, a three-star General. Each of them will put a cover letter summarizing the situation. The three-star will send it to the White House, where they will draft a letter for Nixon's signature. That letter will go to your mother."

"Shit," was all I could manage.

I drafted the letter, delivered it to the Major, who handed it to the Colonel with a smile. The Colonel signed it with a smile and a flourish, and it was on its way.

Later that day, Durnan was back, and suggested we talk over coffee in the Navy cafeteria. He had coffee and pie. I fed the butterflies a can of ginger ale.

"I've done all the background work on this, checked your 201 File over at HQ Company, and confirmed everything you said. There's no doubt your recruiter wasn't up to snuff on this. I've confirmed with the Special Services Office a hockey team doesn't exist. You got recruited under false pretenses." He looked at me for a reaction. If there was one, it was relief.

Durnan continued, "You have options available to you. The first is, we can discharge you from the Army, immediately. Sorry, our mistake. And by immediately, I mean, today."

No doubt about my reaction to that statement— pure joy! It must have shown on my face, or maybe it was my behavior. I think I must have jumped up and shouted something, because I realized I was standing and shouting.

"Before you accept that option, I want you to hear me out. Listen to the other options, and the pluses and minuses of each. The minus on this is the following. After checking your records, I realized you've been in the Army less than six months—one hundred seventy days, to be exact. You're not eligible for the G.I. Bill if you want to go back to school. And, because you haven't served one hundred eighty days, you'll be eligible for the draft again, as soon as they notify your draft board."

"Can't I wait ten days and then decide?"

"I asked about that. Because the action is a fraudulent enlistment, and the Army knows about it today, we have to decide on this option today. Sorry, but today you have to decide if you're in or out."

Seated and silent again, I asked, "What are the other options?"

"You can do nothing, and sit out the remainder of your three-year enlistment right here in the Pentagon. I've talked to the Major and the Colonel, and they're confident you can do the job, so you'd stay right where you are."

"Any other options?"

"There's one more. Within the Army Regulations, there's a section dealing with separations due to fraudulent enlistment, for circumstances deemed by the Secretary of the Army as 'Special,' and that's where this fits. You can't play hockey. You don't want to stay in for the remainder of three years, and you sure as hell don't want to get out tomorrow. 'Special' is the only thing you got left. You have to fill out some paperwork stating the recruiter screwed you with some promises to make you enlist. However, you accept your responsibility to serve your nation, but you don't think they should require you to spend an extra year as a penalty for putting your faith in a less-than-diligent recruiter. Therefore, you'll ask your enlistment be reduced from three years to two years, the same as if you'd been drafted."

"That's it? Will it work? How long will the process take?"

"Yes. Yes. About a month."

"When can I fill out the paperwork?"

"I already filled it out, you just have to sign it and get it over to the company clerk at HQ Company. It's an administrative action that has to start at the company level and go up the chain of command. It'll wind up here in the Pentagon for action, but I've already briefed the action office, and they'll sign off on it. I'll check on it along the way."

I signed the form, thanked Mr. Durnan, and delivered the paperwork to the company clerk after work. The clerk's office was on the ground floor of the Consolidated Barracks where I was still staying. It was a detour of only about one hundred steps to begin my shortcut out of the Army.

The next day, I got a call from the Sergeant Major, not a Command Sergeant Major, just an ordinary one, asking me to come up to HQ Company. The Company Commander wanted to discuss my paperwork regarding the fraudulent enlistment. Mr. Durnan came in later that afternoon to tell me I had an appointment with HQ Company for the next day.

"Yeah. I set it up earlier today."

"Just checking. Told you I'd follow this step by step through the process."

I thanked him, feeling secure I had someone I trusted covering my back.

CHAPTER 23

"I don't Need No Stinkin Special Services"

Even though I made sure I arrived at HQ Company on time, the Sergeant Major made me wait in a stiff wooden chair in the hallway outside his office. The door was open, and I could see him sitting at his desk. A PFC clerk sat across from me at his desk, occasionally looking up, looking at the clock, and then back to his desk.

"Did I get the time wrong," I asked, twenty minutes after the appointed time.

"No, he's just tied up with something," he said, as I watched the Sergeant Major take a shoe shine kit from his desk to touch up his shoes.

Ten minutes and one shoe later, the phone rang. The clerk answered, then pushed the intercom.

"Command Sergeant Major Stoner for you."

The Sergeant Major put his shoe down and answered.

"Stoney, you old dog. God, I haven't talked to you in months. Let's catch up over a few at the NCO Club." A pause while he listened. "Yeah, I've got him cooling his whiney college boy jets outside my office. The boy's supposed to meet with the Major at 1600 hours, but I got him over here early. Teach him a little about how unimportant he is."

Another pause.

"Stoney, this is…"

A longer pause.

The Sergeant Major stood up, the phone in one hand, a shoe in the other. "Yes, Command Sergeant Major. Right away. Sorry to have wasted his time. Yes, I will ask him to call you when he finishes."

With that, the Sergeant Major came to the door and asked me in. Neither of us acknowledged the shoe in his hand, nor did we acknowledge Stoner's call, although he knew I overheard everything. Holding my paperwork, he told me both he and the Major had reviewed the file and had a few questions, he less than the Major. He asked me to summarize what had happened. I did, leaving out the most recent developments precipitated by the letter to President Nixon. He listened attentively while he put on his shoe.

When I finished, he said, "I don't have any questions. Let me check if the Major is ready for you."

What a prick!

With both shoes on, he left the office. When he returned, he told me the Major would see me now.

"Should I call the Command Sergeant Major when I'm done, or will the Major be the one to tell me that?" I had to let him know I overheard his telephone call, and remind him my guns were bigger than his.

Most Army companies have a Captain as the commanding officer. There are exceptions. In Basic Training, my company commander was a First Lieutenant. HQ Company USA proved to be another exception, as we had a Major. Why? Maybe because there were two thousand men in HQ CO, ten times normal. Maybe it was because HQ CO had responsibility for staffing the Pentagon. Whatever the case, I stood in Major John Maxwell's office watching him pace behind his desk.

The Major's office was nothing special, located on the first floor of the Consolidated Barracks. It was the same as my room upstairs, but instead of bunk beds and closet, it had a standard-issue metal desk, and two Army-issue metal filing cabinets. There was also an Army-issue metal conference table, with standard Army-issue metal chairs. Two flags stood in the corner, the Stars and Stripes, and the other I didn't recognize, but it was probably the unit designation. The standard pictures of the President and the Secretary of Defense hung on his wall, along with a couple of guys with stars on their shoulders I didn't recognize. On the top of the filing cabinet he had personal pictures, but I wasn't close enough to make them out. The Major had a window looking out on the water slide, officially designated a loading ramp to the ground level.

"Sit down, West. Got some shrapnel in the 'Nam. The docs didn't get it all out, and it keeps moving around. When a piece gets in a place where they can get at it, they take it out. Still lots to take out," he grimaced.

As I sat down, he kept pacing. If he couldn't sit down, I guess the "'Nam" was his pet name for his ass. Major Maxwell wasn't a big guy, an inch or two shorter than me, making him about five feet seven, and a wiry 140 pounds. His hair was cut short, shorter than the officers in the Pentagon, and turning grey. He wore green fatigues, crisp with starch.

"I've read your file, and quite honestly, I don't understand why you're making this fraudulent enlistment action. After you voluntarily enlisted with a three-year obligation, the Army granted you a delay of three months before you reported for induction. At Fort Dix, you voluntarily changed your enlisted preference from Special Services to Infantry. And now that you have this cushy job at the Pentagon, you want us to forget about all the things in your contract, and let you out early. In the meantime, you get to sit at your desk in the Pentagon and send brave young men off to war, while you laugh at how you beat the system. Is that about right?"

This guy had an agenda different from mine, but I had to keep my cool. Durnan was covering my back, but I still had to get past this guy. I needed him to sign off on the documents before they could go up the chain.

"No, sir. Your use of the word 'voluntary' is not correct. Yes, I enlisted in the Army, but only to avoid the draft. The Induction Center in my home town was sending half the draftees off to Parris Island to become Marines. I did not, I repeat, I did not want to be a Marine." Not that I have anything against the Marines. The stories I heard about Parris Island terrified me.

"That's the only sign of good judgment you've shown."

"The three-month delay in reporting for induction was not a gift, but a formal program of the Army for enlistees, called the Delayed Entry Program. If you asked for it, you got it. No special favors."

The Major said nothing, just looked at me, a look of someone learning something. Why should he know about that? He got soldiers after they finished training, not before they started.

"The change from Special Services to Infantry was because they told me to do so. Personnel at Fort Dix wanted to tear up my original enlistment contract, make it disappear, but I insisted on making the change to the original document. As far as changing from three years to two years, I feel entitled. No reason to serve an extra year because a recruiter either didn't know his job or lied. And finally, sir, I do not send men to Vietnam. I did for a while, and didn't like it. When I asked for a transfer, they could have shipped me out of the Pentagon, but they didn't. Instead, they let me bring men home from Vietnam."

"That's a bit clearer, but you still have other options. If you wanted to be in Special Services, your office might be able to swing that now for you."

"Sir, it's not about being in Special Services, it's about playing hockey, and the Army doesn't have a team. Sir, I don't even want to be in the Army. My preference is to be a civilian again, and being a civilian in two years is a whole lot better than being a civilian in three years."

"I'll have to talk to the Judge Advocate's office about this. I've never done one of these voluntary fraudulent enlistment things before. Mostly, it's involuntary when we use it for the guys who didn't tell us they were homosexuals or criminals when they enlisted."

"Sir, I'm sure the JAG officer will tell you that is only one provision of the regulation. If I may, sir, I can give you the name of someone in the Pentagon who is familiar with this and might answer some of your questions." I gave him Durnan's name and phone number.

"Durnan's not one of the guys in your office, is he?"

"No, sir."

He asked me to wait outside of his office. Unlike the Sergeant Major, he closed the door. Within five minutes he invited me back in.

"Mr. Durnan has become the Pentagon expert on fraudulent enlistments, at least this particular paragraph in the regulations. It also seems a certain three-star general has already told the White House this personnel action will be resolved to your satisfaction. Mr. Durnan asked me if I wanted him to tell the General, who would tell the White House, that you were not satisfied. West, I still have eight years to go to get my twenty. I'd like to go out a Colonel. So, I'll sign your personnel action. Are you satisfied?"

"Yes, sir."

"Good. Make sure you tell Mr. Durnan."

"I'm sure he already knows, sir."

Outside his office, the Sergeant Major waited for me. The walls must be thin, because he also asked me if I was satisfied. When I told him I was, he said, "Good, now you owe me, West."

"Sorry, Sergeant Major, I don't understand."

"I got you in to see the Major. You got what you wanted, so, you owe me."

I paused for a minute. The arrogance of this asshole! "Sergeant Major, I'm new to this 'owing someone' stuff. I'll have to talk with the Command Sergeant Major. I'm sure he'll have some recommendations for how I can pay you back for all you've done." His face morphed into a "Did I just fuck up?" look, as I turned and left.

I found a phone and called the office at 1610. I thanked Stoner for his call, and summarized my talks with the Sergeant Major and the Major. He told me to take the rest of the day off. I thanked him for the twenty-minute vacation, and he signed off saying, "Yeah, Mrs. Stoner says I'm a real peach."

The next day, my official orders arrived, making my assignment in EPCMR-GS permanent. They had been delivered the previous afternoon, while I was over talking to Maxwell. This gave me a better appreciation for what the guys in the field dealt with when they made their personnel requests. It had taken a week, actually six working days, for the paperwork to clear. That might sound like record time for a personnel action, but considering I was the initiator, a week seemed like a long time for paper to go from my desk to my desk.

I told Andy I received the official orders, but he already knew. When I asked if I should tell the Colonel and the Major, he said they knew. I asked if there was anybody who didn't know, like maybe my mother. He said she knew. Durnan called her.

"So, I'm the last to find out?"

"Not the last. There's still lots of people who don't know. Just in China alone there are about a billion people. They don't know, and they probably don't give a shit. It's not all about you, you know."

I now worked at the stuff on my desk with a different outlook. This is what I'd be doing for the next year and a half, most of that time without Andy showing me the way. I harbored significant doubts about my ability to do the job as well as he did, despite my "better than Andy" rating by Durnan. I was not in Andy's league.

My desk felt like my desk now. It faced Andy's, permitting me to see only a small portion of the larger office. The office was two large rectangles. My desk sat at the corner where the two rectangles met, looking into the smaller one. The bulk of the office, its people and activity, were behind me, including the desks of the Major, the Colonel, the Sergeant Major, and Julie. I could still smell Julie's perfume, but I couldn't see the back of her neck anymore. The piece of the office I could see, I studied closely.

All the furniture looked like someone had bought it at a government used furniture auction, at the end of the day on the last day of a sale. This collection of junk was made more unsightly by being arranged haphazardly. They jammed desks and chairs wherever they fit. While all the desks were metal, they were all different. The same with the chairs. There were no pictures on the wall, the only exception being the wall behind the area shared by the Major and the Colonel. Like Maxwell's office yesterday, the obligatory pictures of the President, and the Secretary of Defense hung. If every office had to have these pictures, I wanted to be the guy with the contract for supplying them.

Except for the rare small picture of a child on a civilian desk, there were few personal effects. None of the soldiers had pictures. On the wall next to some soldier's desk hung his ETS calendar. On another wall, the outline of the cinder blocks was barely visible under the layers of paint. There were enough layers of paint on the walls to shrink the elbow room, probably a desk's worth of space.

Each desk sported a black plastic telephone with multiple lines. The only modern office conveniences were the new IBM Selectric typewriters, with the magic balls that whirled around. One desk I now faced had a soldier whom I hadn't met yet. Andy told me he was a Mormon who kept to himself. Friendly enough guy, but he was put off with the language some guys used, as well as some college-like humor. The word was, he typed faster than the Selectric was designed to handle. The IBM folks had come in to test him, and sure enough, he beat up their machine. He took dictation on the machine faster than a stenographer could stenog. I was close enough that when he typed, I couldn't hear the individual keystrokes, just a "brrrrrrrr."

He and I enjoyed the best view in the office. His desk was in front of mine, behind Andy, but at a right angle to us, with his back to the wall. He was deeper into the smaller room, so he had less of the office to view, but the view he had couldn't be beat. His desk butted up to and faced Pam.

Pam was the most beautiful woman I have ever seen. The first thing that got your attention was her body. On a scale of one to ten, she pinned the needle at twelve. The person who dreamed up the Barbie doll must have used her body as a model. She wore the same flowered print dresses the other women in the office wore, but while theirs were formless and hung on their bodies, hers seemed to be painted on above the waist, and the paint rolled down her hips, then dropped to her knees as a skirt. She always wore high heels, which emphasized her legs. When she stood, she posed her body, not in a sexy pose, but a pose that was sexy. She had a stance like a proud race horse, legs thrust back under her body. She had short, shiny black hair. Her eyes sat above high cheek bones, her nose flared just a little, and her lips were full, always emphasized by red lipstick. She adhered to the rules laid out by Mrs. Ramirez for proper attire and decorum, but the rules didn't matter with her. If they changed the rules to make her wear one of those Arab things that hide everything but her eyes, she still would have been an eleven.

Every time she walked past my desk, I stopped what I was doing and followed her out of the corner of my eye. Pam wasn't teasing, but she knew every guy in the room watched her and wanted her.

"The bad news is, she will never give you the time of day, but she'll be friendly." Andy said this as he stared at the paperwork on his desk and I stared at Pam as she moved away from me, back to her desk.

"Is there good news?"

"Yes. You'll get over it quickly."

"How so?"

"That big Major from White House Liaison is Mr. Pam."

"Oh. I'm over it already. Major Mr. Pam is Mr. Durnan's boss."

When I told Andy about my interview at HQ CO, and the "owe-sie" claimed by the Sergeant Major, he rolled his eyes and shrugged, explaining there were a few guys in the Army who were playing fast and loose.

"They know the rules," he said. "But they choose to apply them to their own benefit. Like your recruiter. He and the Sergeant Major up at HQ CO seem to be cut from the same cloth."

"Yeah, I'd like to give that recruiter an 'owe-sie' of my own."

Andy put down his pen and looked at me. He didn't say anything, but I could see he was thinking. "Have you had any contact with the recruiter since signing up?" I told him about the call I made from Fort Dix, when they forced me to change my enlistment contract. Andy smiled when I told him the recruiter told me to "stop your fucking whining. You're in the Army now."

"Let's call him and tell him the good news about your new assignment, and how happy you are now." Andy rehearsed me for my "happy" talk.

I placed the call to the recruiting office, and he answered on the second ring.

"This is PFC West. Remember me? You recruited me to play hockey in Special Services. How are you doing?"

"What do you want this time, West?"

"Our last call didn't end on a good note, and I just wanted to call and tell you I've got my permanent assignment and I'm very happy."

"Where are you?"

"I'm a clerk and just got my orders to DA, EPCMR-GS."

The line went silent for a few seconds. "West, I don't want to burst your bubble, but you're reading the wrong line of your orders. EPCMR-GS at DA is the office at the Pentagon that makes all the assignments for clerks. I know because it's the office that makes my assignment as a recruiter. You're reading the authorizing line. You have to read the 'To' line to find out where you're going."

"Sergeant, the authorizing line and the 'To' line read the same."

"Where are you calling from, West?"

"I'm calling from my new desk in the Pentagon, from EPCMR-GS. And Sergeant, you're ab-so-fucking-lute-ly right, it is the office that handles recruiters. In fact, I've got your IBM card in my hand right now. It says you've been at your assignment for over two-and-a-half years of a three-year assignment. You'll be eligible for reassignment in six months."

Nervous silence reigned on his end.

"Sergeant, would you like me to see if I can get you an assignment as a recruiter in the Special Services Branch?"

Andy smiled at my ad lib.

"There's no such thing as a recruiter for the Special Services Branch," he said cautiously.

"Right, just like there was no hockey team in Special Services when you got me to sign up."

"Hey, buddy. Let's not—"

I interrupted him saying, "Sergeant, I have to get back to work. So many assignments to make, and so little time to do it. Good talking to you. I'll keep an eye open for you when it's time for you to rotate."

"West—" was all I heard, as I hung up.

"That was fun. It'll make him think for a while."

Andy picked up his pen and started working again. "The Sergeant's going to think you've already got him on orders to Vietnam."

"We can't do that, can we?"

"No. We can send the recruiters to recruiter stations in the States, and we can pull them out of Vietnam, but it's the IBM machine does the selection when it's time for them to rotate overseas. We can still watch him, though."

"I don't like the man for what he did to me, but I'd feel terrible if I sent him to Vietnam and he got hurt or worse. I didn't like sending the AIT guys with kids to Vietnam. Just because this guy's a prick wouldn't make it any easier to live with."

"Agreed, but it doesn't mean we can't have a little fun with this guy."

221

CHAPTER 24

"Driving Mrs. Stoner's husband"

A furnished apartment became available in mid-November. Les decided to share it with me, saying he wanted a place away from the military, and we moved in over the weekend. Les had a car, but his social schedule on weekends guaranteed I'd be sitting at home every weekend. So, one thing on my list was to look for a car.

On Monday morning, Sergeant Major picked us up, as agreed. Ray sat in the back seat. As I moved to the passenger side, the Sergeant Major emerged from behind the wheel, telling me I would be driving.

"Ray makes me nauseous when he drives, and when I drive in traffic, I get nervous and jerky. So, you drive." Looking at Les, he said, "You're in the back seat with Ray."

The car was utilitarian, like the owner—a brown, four-door Chevy. I adjusted the seat and the rear-view mirror before backing up.

"Stay off Shirley Highway. Take South Walter Reed Drive to Columbia Pike. Take a right at the Big Boy, and go in the entrance to South Post, just after Henderson Hall."

I took a quick look at him, but before I could speak, he said, "Eyes forward. Henderson Hall is the Marine Barracks. The Marines didn't want to be part of the Consolidated designation at Fort Myer. I figure they didn't want to play second fiddle to the Third Infantry."

"I don't know where that entrance is," I said.

"Don't worry, I'll tell you," to which Ray cleared his throat in the back seat.

"In addition to making me sick, that's why Ray is in the back seat."

From the back seat, Ray added, "Ray is happy to be in the back seat, and happy someone else is driving. Maybe you won't be nervous anymore."

"Do you mean, I'll always be jerky?"

Les had to chime in, "What a happy group of fellas this is."

"If you'd like, we can let you out."

"No thanks, Sergeant Major. I think this is a great way to start my day. It'll make me happy to get to work."

Another quick peek toward the Sergeant Major and I saw him smiling. Seems he liked Les. "Good answer, Private."

When we got to South Post, I parked the car and handed the keys to the Sergeant Major.

"Keep them, because you're driving home. Don't lose 'em."

Andy and the rest of the guys wanted to know how the ride in went, now that I was Stoner's new driver.

"Okay, but it was only the first day. He made me nervous."

"That's understandable, but it'll be fine."

The ride home proved uneventful. As we approached his neighborhood, he told me to drop him and Ray off, take the car for the night, and pick him up in the morning.

The commute followed the same pattern for the rest of the week. On Friday, when I asked if he wanted me to go to the apartment first, he said no.

"Mrs. Stoner has a car, and on the weekends, we spend a lot of time together. The only time I need this car is to and from work. Use it if you need it."

I didn't know what to say, but Les did. "Should he wash it, fill up the tank, and get it pretty for next week?"

"No need to wash it. When the tank gets half empty, we fill it up. I filled it up last time, so it's Ray's turn next. You can fill it up, but he pays. After that, you two decide who's next." With that, he was out of the car, with a "Have a nice weekend."

A month later, the week before Christmas, the official response to my request to revert to the remainder of a two-year enlistment came back. It wasn't the approval I expected; it was a request for more information. When Personnel at Fort Benjamin Harrison, the Administrative Center of the Army, checked its records, they couldn't find me. They couldn't verify I was in the Army. So, I had to take my IBM card out from under my blotter and put it in the system for a month—so when the guy who was checking off that box checked the records, they could say, "Yes, he's in the Army."

It wasn't the quickest process, but it was the easiest. Once the first request for more information came back, Mr. Durnan took over. I never saw it again until they approved it. He came in regularly and gave me an update. He explained that particular box had to be checked, because with the deaths increasing in Vietnam, the Army was being very careful on personnel actions like this. They didn't want to make a mistake if I was already dead. Not having an IBM card in the system being one of the indicators of being dead. It took a month for the information to get back into the system, and another month until I got the approved paperwork.

Now, I became one of the few men in the Army who had an IBM card coded as being a two-year enlistee. Others would first learn of it that wonderful day when I crossed off three hundred sixty-six days from my ETS calendar, the extra day because my third year, 1972, had been a leap year.

The end of 1969 brought other changes. In September, Nixon had ordered the withdrawal of 75,000 American troops from Vietnam, in the first phase of Vietnamization of the war. This was followed in December with the second phase of withdrawal, where another 50,000 American troops were withdrawn. Between these two events, peace protesters descended on Washington and other major cities. The Moratorium Demonstration, on October 15, was minor compared to the Mobilization Demonstration the following month, which drew over 500,000 protesters to Washington. The situation was so tense for the second demonstration, the Secretary of Defense ordered all military personnel working at the Pentagon to stay home, and brought in the 82nd Airborne Division from Fort Bragg to provide security for the building. By Christmas, things settled down.

CHAPTER 25

"Less Than Stone"

Les and I settled into the apartment in Arlington. The apartment had an Arlington address, but part of it sat in Alexandria. Les's bedroom was in Arlington, but the rest of the apartment was in Alexandria. The office for the apartment complex was directly below his bedroom, so the complex had the Arlington address. A unique piece of history and geography.

When they drew up the original boundaries of the District of Columbia, it formed a perfect square, ten miles by ten miles, taking parts of Virginia and Maryland. The planners rotated the square on a map, so it formed a diamond. With time, Virginia got back its piece of the District. When Arlington and Alexandria decided on the boundary between their two cities, they chose the original District line.

After we moved in, we settled into a routine. When Les was home, he fought the alarm for every second of sleep. Being the first up, I'd shave, shower, dress, and have the coffee going before he got up. Many times, I'd grab my coffee and go outside to enjoy a cigarette. Those mornings, the first human contact I would have would be the grouchy old lady across the parking lot, waiting for her dog to pee on someone's car tire.

The ride in to the parking lot at Fort Myer was chatty. I had been up for a while and had a cup of coffee. At one point, I suggested a twenty-minute earlier start, so we could have breakfast at the South Post mess.

The Sergeant Major rejected my proposal. "Breakfast at that mess hall is not something you rush through in twenty minutes. The line cooks take time to make you a nice meal. So, you should respect them by taking time to eat it. Second, you don't need breakfast every day. In the short time you've been here, I can see you've put on weight. Breakfast means more weight. We don't need another Teddy who doesn't look like a soldier."

"Then I'll need to stop for coffee on the way. One cup isn't enough to start the day."

"There's no place to stop for coffee without taking us out of the way. Taking us out of the way, makes our trip longer. How do you like your coffee?" Stoner asked looking straight ahead.

"Cream and sugar."

"Drinking it black is better for you; fewer calories. Besides, where would you get cream and sugar when you're hunkered down on a patrol in a combat zone?"

"If I'm ever hunkered down in a combat zone, I won't be thinking about cream and sugar for my coffee. I won't even be thinking about coffee. More likely, I'll be thinking about toilet paper."

"Fair enough," he said, with just the least bit of a twitch of his eye.

The next morning, Stoner got in the car with a small thermos of coffee. After he unscrewed the top to the wide-mouth bottle, he poured a small portion into the cap and handed me the thermos. "Black," he said, "just the way I like it."

From the back seat, Ray asked, "Sergeant Major. Didn't your mother ever tell you when you share, you should bring enough to share with everyone?"

"Yes, she did, but this isn't sharing. This is protecting the mission. The mission is to get safely to work. Can't do that if we have our driver grouchy and sleepy in this traffic."

"Are you going to bring him coffee every morning?" I could see Ray in the rear-view mirror, a smile on his face, testing Stoner.

"Glad you asked. No, I'm not. West will be grateful for this gesture. He will return the thermos on the ride home, rinsed clean. He will give it to you, and tomorrow morning, you will fill it with coffee, enough for him and me, to protect the safety of our mission. We will alternate each day."

"What about Les?"

The Sergeant Major turned toward the back of the car to see a sleeping Les, mouth open, snoring. "I don't want to drink any coffee he makes this early in the morning."

"Understood."

"Cream and sugar, please," I said to Ray.

"Black," corrected Stoner. And that's the way it was for the next year and a half. One day I would get a thermos of black coffee from Stoner, and the next day, I would get it from Ray. The mission stayed safe, no accidents, no delays. Once a day, I drank black coffee, thinking of the guys in Vietnam.

Stoner and Ray both enjoyed history, Ray by education, Stoner by experience and a quest for knowledge of the things around him. Ray studied World War II in graduate school, with particular interest in the Reichstag Fire. Stoner had fought in WWII, first as a brown-shoe enlistee in the Canadian Army who lied about his age to gain entry, and then as a G.I. in Europe, and later, the Pacific. Ray read about the war; Stoner lived the war. The discussions were lively, but I seldom contributed to the conversation, content to listen and learn.

During these rides to and from work, I got an insight into Stoner. He didn't brag, but gave us bits and pieces of his background, when the occasion arose. One morning, the subject of Elvis Presley came up. He was making a comeback against the Beatles, who had taken his throne. Elvis remade himself and was now all aglitter, in a manlier version of the Liberace style.

"He was a nice young man. I gave him a lot of credit when he reported for induction. When he got drafted, he didn't ask for any special favors. Would have been easy for him to get a Special Services assignment that allowed him to tour and give shows to the G.I.s, but he chose to be a truck driver. He was courteous and respectful. I think his parents had a lot to do with that."

"Did you serve on a post where he was stationed?" I asked.

"No, I was stateside the whole time of his service. But I was stationed at the Memphis recruiting station when he showed up in 1958. I ushered his group through the swearing-in, and then through the medical tests at the VA Hospital. I put them on the bus to Fort Chaffee. What a circus! All the reporters and girls trying to get a last picture or wave. Throughout it all, he kept his cool. As soon as he took that step forward, he accepted the Army and called everyone 'Sir.' I always remember him each year on that day, March 24, and his service number has stayed in my head all these years, 5331-0761. Don't know why, it just does." Stoner said it as a matter-of-fact, not bragging at being close to stardom.

The Sergeant Major didn't share much about his other assignments. He did say his deployment to the Panama Canal Zone for three years provided a high point for his family. The Stoner children, in their early teens, had a wonderful time. He got home most nights and weekends, and the family enjoyed a normal life. The only exception to a normal life was providing a housekeeper for his wife. Mrs. Stoner enjoyed life in the tropics without the drudgery of cooking and cleaning unless she wanted to, and she seldom wanted to.

Walking as a group with Stoner between the South Post parking lot and the Pentagon was memorable, beginning with the very first day.

When an enlisted man passed an officer outside, he was supposed to salute. Long before I arrived at the Pentagon, this simple sign of respect became a burden for the officers. The officers arriving in the morning would have to pass a solid stream of Army and Air Force enlisted men and women who worked the night shift. To return every salute meant the officers would walk with their hand at their hat for the entire quarter mile, as would the enlisted men. To avoid this, the officers walked with a sideways glance. The soldiers saw this and picked up on it, and looked the other way. A good compromise—neither had to be constantly saluting. There was the occasional visitor or high-ranking officer, and then things returned to normal.

That first walk with Stoner had two breaches of military courtesy. The first, when Ray tried to walk on Stoner's left. Stoner tried to maneuver to the left, but Ray blocked him. Stoner stopped us dead.

"The senior officer or NCO always walks on the left. Don't you know anything?" He took his spot on the left. We had been walking on the right side of the road, with the traffic to our left.

Ray tried to have fun with Stoner. "Oh, now I get it. Like we're on a date. You walk on the side with the traffic, to protect us from a car that might jump the curb. I remember my mom telling me that when I started dating."

Not to be outdone, I said, "My mom told me the same thing. But I always wondered about that. It started a long time ago in England. I would have thought the man would walk on the inside in case someone emptied the bed pan out a window."

Stoner didn't miss a step or react to either of us with any humor at all. "It's a tradition from when soldiers carried swords. The most senior person always walked on the far left. Do you know why?"

"So he wouldn't get the bed pan on his uniform?"

"No. Because if he was walking with two clowns like you, he wouldn't have his head cut off if you had to draw your sword." He said this while demonstrating the action of drawing a sword, knocking Ray's hat off as he did.

Ray picked up his hat and said, "I don't carry a sword, and I'm glad you don't, either."

He had made his point.

The next point was almost a disaster. A First Lieutenant approached us. Stoner saluted. The Lieutenant responded with a surprised salute. We didn't salute! The Lieutenant stopped and lit into us.

"You're supposed to salute an officer when outdoors. Didn't they teach you that in Basic Training?"

Stoner saluted again, saying, "Sorry, sir. Have a nice day, sir."

Ray took the first step off the cliff. "An officer? Sorry, I didn't see an officer. Did you see an officer, Jonathan?"

I took one step off the same cliff. "Nope. I know officers. They're the guys who have that fancy gold stuff on their hats. Scrambled eggs, they call it."

Stoner stopped us. "That was a First Lieutenant, an officer."

Ray didn't recognize the serious tone in Stoner's voice, or else chose to ignore it. "Sorry, I don't believe I've seen a Lieutenant since I've been here. I remember now—they have one up in JCS who answers the phone and makes coffee. I guess I just forgot they were officers."

"Don't forget Lieutenants are officers, or I'll remember you have your IBM card under your blotter. Show them the respect they've earned and deserve."

That was it. Our new routine was to walk to the right of the Sergeant Major, and salute all the officers. A lot of the officers didn't like the change, but screw 'em. They didn't know where our IBM cards were.

Despite being the Sergeant Major's daily driver, I received no favorable treatment from him. More accurately, I received no more favorable treatment from him than any enlisted man in the office. He told us he felt privileged to manage us. I imagine during his career commanding enlisted men he had never been responsible for a more educated group. The average education for an enlisted man in our office was two years beyond a Bachelor's degree. He valued education and had insisted his own children get at least a Bachelor's. He himself had taken all the Army courses offered that helped him understand his job. Additionally, he had over one hundred fifty credit hours, through the Army's college education program with the University of Maryland. While he had more than enough hours for a degree, he didn't have one. When I asked him once, he said he didn't need a degree. What he wanted was knowledge, and that's why he took the courses.

Stoner was a voracious reader, mostly history and biography. I don't know how much time he spent at home reading, but during the work day, he spent exactly twenty-five minutes each day reading during his lunch hour. The allotted time for lunch was forty-five minutes. The Sergeant Major always brought his lunch, usually a half sandwich and a piece of fruit, which he ate at his desk with a cup of coffee in ten minutes. His meal was spartan and could have been eaten much faster, but he allowed himself the luxury of the ten minutes. After eating, he would check his calendar, write something on the back of his hand, and then take the five-minute walk to the Concourse of the Pentagon, the massive shopping area that provided virtually everything the 35,000 employees of the building could need. These were the things taken for granted by workers in the cities and towns around the country: food courts, a bank, dry cleaners, drugstore, and a book store, Brentano's, Stoner's destination each day.

I only learned of his regimented routine one day when he came back to the office twenty minutes early. I asked if everything was okay.

"My book was gone."

"You went to the library?" In addition to everything else, the Pentagon has a large library, the JAG library. Most of it supports the Judge Advocate General staff, but over time, the collection has expanded to include a broader collection of literature.

"No, Brentano's."

"I'm sorry, I don't understand. Your book was gone?"

"No, they sold my book," he said with emphasis on "my."

"You wrote a book?"

It was the only time he ever gave me a look that made me feel dumb. His eyes rolled up, and he exhaled in exasperation, a sigh that sounded like "you fucking moron" in Army senior NCO-ese.

"No, I read a book at Brentano's. Every day I go to Brentano's and read a few chapters of a new book. I write the last page I read on the palm of my hand, I come back, write the page number on tomorrow's calendar. The next day, I write the page I left off on, on the palm of my hand, I go back to Brentano's, start at that page, read for twenty-five minutes, and come back. I repeat that process every day. Today, the book was gone. They sold my fucking book."

"Why don't you just buy the book?"

"I have too many books already. I read them once, remember what's in them, and never read them again. Mrs. Stoner complains every time we move, we have more boxes of books than anything else." Pausing, a smile forming, he said, "She's a sweet lady, you'll have to meet her sometime." Then the curmudgeon returned. "So, I don't buy books anymore. I use the library, or I use Brentano's."

"So, get the book from the library."

"I tried, but it's a new book, and I'm on the waiting list. Might be three months before I get it."

Stoner remained grouchy for the rest of the day, including the ride home. The mood extended to the next day. Remaining at his desk after eating his lunch, he looked like a man whose dog died. The other guys in the office noticed and asked. I told them about his routine being put off by someone buying the book he was reading at Brentano's.

"What was he reading?" Andy asked.

"I don't know. Should I find out?"

"Wouldn't hurt, someone may have it. We all read a lot."

I asked Stoner. He told me he had become a fan of Alexander Solzhenitsyn after reading First Circle during his last tour in Vietnam. The Russian's writing soothed him. Now he was reading his latest novel, Cancer Ward, the story inspired by the author's mother's fight with cancer. The novel gave an insight into another aspect of the social system of the Soviet Union, something that currently interested him. He felt the social upheaval in the United States, the feeling of entitlement that was growing in several areas could lead to aspects of socialism creeping into our culture. The creep would not be called or recognized as communism, but if left unchecked, it would ruin our country.

"Cancer Ward, by some Russian," I told Andy.

"No shit. My wife just finished it. It was a highlighted selection in the book club she belongs to. She loved it. I was going to start it this weekend. Can I drop it off at your place tonight, so you can give it to him in the morning?"

"Why don't you just bring it in tomorrow?"

"I start a long weekend tonight. Going to New York for a wedding. You give it to him."

The following morning, when Stoner gave me my coffee, I gave him the book. "It's not a gift, it's a loan. Belongs to Andy's wife. Andy wants to read it when you're done, and I'm going to read it after him."

Stoner just held the book in his hands and looked at it. He mumbled a thanks, his eyes glassy, like he was tearing up. He didn't say anything more for the rest of the ride.

When we got to the office, his mood changed back to the non-grouchy lifer we had been used to. The change in his behavior back to normal was noticed and appreciated. Only Andy, Ray, and I knew why, and we kept it that way. Mullaney announced everything was "back to abnormal."

Months later, I met Mrs. Stoner for the first time.

The occasion was an invitation to dinner with her and the Sergeant Major. It took her weeks to get Stoner to agree to have me over, because he had a rule that no one in his command ever interacted with his family, except at official military functions. She convinced him I wasn't really in the Army, just a nice kid working in his office for a couple of years who had to wear a uniform. He gave in. Stoner was right, she was a very nice lady and made me feel welcome and comfortable in her home. She pulled me aside after sending Stoner into the yard to tend to the meat on the grill. She told me he had been really moved by the effort I had gone through to get him the book. He had always been stand-offish with his men, rarely socialized with them. The only times his men had ever done anything nice for him was when dictated by protocol, such as a promotion, or an assignment change. She said he often joked that the only time he got a party or gift from his men was when he left.

So, under the crusty exterior of one of the highest-ranking enlisted men in the Army—a crust that began to form in the brown-shoe Army of the 1930s—there was a man who appreciated being cared about. Stoner's heart was less than stone.

I remained friends with the Stoners for the rest of my time in the Washington area, before moving on with the rest of my life. We remained friends after he retired, visiting a couple of times. The only thing I remember changing with time was Stoner smiled a little more.

CHAPTER 26

"Diplomacy"

Most people don't associate the Pentagon with diplomacy. They think of the State Department. The Pentagon with its war machine is where the President goes when diplomacy fails. Not so for those of us fortunate enough to be here hiding our IBM cards under our desk blotters. We considered Diplomacy a war game, with a capital D.

It all began Christmas of 1969, when Andy received a gift from his wife. Being a law student, or former law student, or law student taking a two-year sabbatical to study the art of war, he had a natural interest in all things historical, and Diplomacy fit the bill. At least, his wife thought so.

Diplomacy was a board game, invented in the 1950s by a guy at Harvard. He made a game of the events leading up to and ending with World War I, i.e., when diplomacy failed.

Andy found playing it with his wife unsatisfactory, for several reasons. They designed the game for seven players, all representing a country in the early twentieth century: England, France, Germany, Italy, Austria-Hungary, Russia, and Turkey. He soon learned it wasn't much fun playing a game intended for seven people with only two. Equally important, because it involved lying and deceit, it was hard to

238

play the game with his wife and get laid afterward. He invited six guys from the office to an evening at his house to play the game. I guess he thought we were used to being lied to, deceived, and not getting laid. He further enticed us by saying he would supply beer and pizza.

Andy set the game for a Saturday night at six-thirty. The seven included Andy, me, Mullaney (another law student on sabbatical), Ray (history graduate student on sabbatical), Peter, Bill (an insurance underwriter on sabbatical), and Les.

Andy lived with his wife in an apartment complex built during the Korean War era, when the government was growing. With the District run down, living across the river in Alexandria and Arlington presented a way to be safe, secure, and near like-minded people who worked for the government. Most apartment complexes had a history of a government celebrity who had lived there. The one I lived in claimed to have been the temporary home of the Nixons when they first came to Washington. Andy's complex claimed no former celebrity. He said in the years to come, they would claim him as the celebrity. The apartment was nicely furnished, probably by his working wife, in Danish modern—stylish, not the overstuffed things we grew up with, and cheap, because it was sticks and pads. Both of these attributes contributed to the demise of Danish modern, because sticks and pads proved to be uncomfortable.

The apartment was a typical size for the area, about eight hundred square feet. It had a living room with a dining area connected, but set aside. A small galley kitchen, two bedrooms, and a small bathroom completed the floor plan. A door to one bedroom was closed, the retreat of his wife this evening. The other bedroom he called the study, a bedroom converted to an office. The study was a relic from college dorm days, with a desk made from a door, and book shelves made from cinder blocks and planks.

Andy explained the rules. He modified them a bit because it made for an easier and more interesting game. The inventor had provided for the game to begin in 1901, but Andy wanted it to start in 1914.

"Why?" I asked.

"Because I think it's better. The game is complicated enough, and takes a long time to play. This change should make it shorter and more focused."

Ray jumped in. "How fucking complicated can it be? It's WWI. Everyone knows how it turns out."

"Please. Just listen. Each of us will be a country," he started.

Bill said, "If this is WWI, I want to be the United States."

"The U.S. isn't in the game."

Several people jumped in, "How can it be WWI if the U.S. isn't in the game?"

"Because the game starts before the U.S. gets involved."

Bill wouldn't let up. "I'll sit out until 1917, and jump in then as the U.S."

Andy looked toward him and continued. "Please, just listen," he pleaded. "Each of us will be a country: England, France, Germany, Italy, Austria-Hungary, Russia, and Turkey."

"You let Turkey in the game, but you don't let the U.S.? Now I know why it's complicated."

"Please just listen. I didn't make up the rules—"

"Yes, you did. You just said you changed the rules by starting in 1914 instead of 1901. Why don't you change them a little more, and get rid of Turkey and add the U.S.?"

"Maybe next game, after we learn how to play."

Ray took his shot again. "Can't be very hard to learn, man. It's WWI. We know how it turns out. Germany loses, and Americans learn how to French kiss and get blow jobs."

Everybody laughed except Andy. "Please listen."

"Okay, okay. It's your game, your house, your beer, and your pizza. Talk."

"Each of us is a country. Before you ask, you can't pick who you want to be, we draw for it."

Les had an imaginary pencil in the air. "Betcha everyone draws Italy, 'cuz it's easy to draw. Looks like a boot. And no one draws Austria-Hungary, 'cuz they don't know what the fuck it looks like."

"Everybody knows what a turkey looks like," I said, continuing Les's imaginary air drawing with an air bird. "Could be that's why Turkey is in the fucking game, and the U.S. isn't."

Everybody laughed again, except Andy. He was getting pissed, and the evening could come to an early end before the beer and pizza. "Maybe we can pretend we're not in the Army, and we're the well-educated people we were before we got into the Army, and avoid using 'fuck' for the rest of the night," he said.

When I agreed by saying, "Fucking A," Andy sent me a look of disapproval.

Bill jumped in and saved me before I dug the hole any deeper. "Sure, no problem. If we can pretend the U.S. wasn't in WWI, we can try to pretend we're educated."

Finally, Andy explained the rules to us. Each of us would be one of the seven countries, referred to as the Great Powers by the inventor. Andy preferred the term "country." Each country would have three tokens, representing either an army or a navy. For example, with England, because of its geography as an island, each of the tokens would be a navy. Austria-Hungary, being land-locked, would have three armies. Other countries, because they had a coast border and a land border, would have a combination of armies and navies, but there could be only three per country. Each country was divided into three different areas.

As a country, we would negotiate with other countries to make alliances. These alliances were for the purpose of defending against aggression by another country, or for making an aggressive act against a country. We had thirty minutes to have our discussions between each move. Each move represented six months of real time of WWI. Once the thirty minutes were up, we would gather around the game board, which was a map of Europe in 1914, and read the moves.

It was a simple sum-total game. For example, if Italy and Germany formed an alliance, with Italy agreeing to support an aggressive act against France, then the move would be two against one, Germany and Italy against France. France would lose, and Germany would occupy a piece of France. France would lose a token and Germany would gain a token. Once you lost all your tokens, you're out of the game, and your country has been overrun. He cautioned us that while Italy or Germany might attack France, Russia or Austria-Hungary could attack them, so a token would have to be used to defend.

"Sounds easy enough. Why do we need thirty minutes for a move? Everybody knows how it will turn out, so let's do ten minutes," Ray suggested, as he looked toward the counter with the snacks. "That way, we can get to the pizza and beer sooner."

"Let's stick with the thirty minutes for a start. If we need to change, we can."

After agreeing, we drew for countries. Ray got Germany. At first delighted because of his interest in German history, then he realized Germany was the target of the rest of us, and became pensive. He revived and accepted the challenge to rewrite history. I got England and her three navies. In the first move, I would be relegated to a support mode, because none of my borders touched the border of another country. Support and defense of the English Channel was my only available first move. We hadn't even started, and the geography of the board map had already limited my actions, just as geopolitics had limited England early in WWI. I might like this game, I thought.

We paired off to begin discussions. It's hard to tell how much of what we knew about WWI was dictating our alliances. Everybody wanted to talk to England, France, and Austria-Hungary. Nobody wanted to talk to Russia or Turkey, forcing them to talk to themselves. Nobody wanted to talk to Germany or Italy, forcing them to talk to themselves. Andy called time at thirty minutes, but we all asked for more time.

"I've got to talk to England," said Austria-Hungary.

"I've got to talk to France again."

That was something we hadn't anticipated. After you had talked to someone, you assumed it was over. We learned quickly that, like real diplomacy, events and agreements often changed earlier agreements.

To avoid being perceived as a liar, it sometimes became necessary to go back and cover tracks. This happened to everyone. Andy called time after another thirty minutes, and was met with the same chorus of "more time."

And so it went for the rest of the evening. At ten o'clock, after having eaten cold pizza, after having drunk all his beer and stunk up his house with cigarette smoke, he called for a move. "If you don't have moves to submit, I will consider you only have defensive positions, no aggression, no support. Two minutes for your moves."

We had spent three-and-a-half hours, and had yet to make a move. We were prompted in this decision each time Andy's wife came out, asked how it was going, only to be told we hadn't made the first move yet. The second time she came out, her reply of "The best the Army has in the clerk schools" told us what she thought of us.

We made our moves and assembled around the game board. As Andy read each move, Peter put the pieces in their correct places. Before making a final action, Andy would ask if there were any other moves impacting those taking place. There were, so all the moves had to be read, and then the sum-total evaluated. Then, and only then, could the fate of the tokens be determined.

So, at a minimum, we had seven countries, each making three moves, all of which had to arbitrated with any other moves being made. After thirty minutes, all the moves were recorded, adjudicated by Andy, and set for the next move. In the final tally, France lost one token, an army, and everything else remained in place. I gained a slight advantage, as I was able to move one of my navies into the English Channel and another into the North Sea. These would be good defensive moves, and if necessary, allow me to attack Germany, France, or Russia in the future.

Another move that night was not in the cards. Andy told us to go home. He summarized the evening as a practice game, to get us familiar with the rules and how to play. After agreeing we wanted to play again, he told us the next time we would have to stick with the time limits, or we would never finish a game, and we had to bring our own drinks and contribute to the pizza fund. We agreed. He also said he might look at the rules and tweak them, to make the game move faster.

"If we have to learn new rules, won't that slow things down?" Bill asked.

"Let's wait until I have something for you to bitch about before you start bitching, okay?"

After several weeks, Andy invited us to his house for an all-day Diplomacy game on a Saturday. His wife would be out of town, visiting family, and we might just have enough time to finish a game. We agreed to a 1:00 p.m. start, leaving us the morning to take care of any chores or family obligations. Some guys learned their spontaneous agreement to spend most of a weekend day away from their family was not a decision they should have made on their own.

The game started on time, with no new rules from Andy. Our previous experience did little to speed up the deliberations between moves. First moves took three hours and the adjudication another fifteen minutes—an improvement, but well outside the thirty minutes suggested. Second moves, with adjudication, had the time down to two hours and forty minutes. The third move was completed at eight o'clock. We had been at the game for seven hours, made three moves, and had only succeeded in removing two French tokens and one from Austria-Hungary. Germany was advancing, and England was not in a position to do anything yet. We had spent our day doing exactly what history had recorded over fifty years earlier.

We would never finish this game in one sitting, and some couldn't afford to take another day from family, home life, or part-time jobs. It looked like the end of the game for us. We parted, thanking Andy for his hospitality, sharing a sense of group failure, and some dreading the cool receptions awaiting them at home.

On Monday, Peter asked if the game players could meet for five minutes at the end of the day. When we assembled, we had the office to ourselves, the civilians having made a beeline out the door at precisely five o'clock, as had the Major and the Colonel. Stoner sat at his desk, waiting for his ride, but out of the range of our discussion.

"I got thinking about the game yesterday, while I was watching my girls coloring." Peter said this as he held up a map of the United States he had taken from a coloring book. "It occurred to me we could still play Diplomacy with just a minor change."

Andy gave Peter a curious, challenging look. "What's the minor change?"

"Well, we're never going to finish a game if we try to do it at one time at Andy's house. But we all work together five days a week. We're here eight hours a day. We interact with each other on lots of issues, and we socialize over coffee and at lunch. I got to thinking we'd have time during the week to do all our 'diplomacing' right here. Then, once a week, we could stay for a few minutes and let Andy referee the moves. Then we could start another move the next day."

"How are we going to keep track of the board?"

"Sorry, that was the critical piece, and I forget to tell you. I could make a Xerox copy of the board. I'd let my daughters color all the countries a different color. We'd each have a copy to work from during the week. Then, when Andy finishes the refereeing, I'd mark a blank copy, take it home, and let my girls color seven copies."

Ray puffed up. "Why can't we color our own copies?"

It's hard to say whether he was serious, but each of us looked at him with an "Are you serious?" look.

"Just asking. It would be okay with me if Peter's kids did the coloring."

"What about Bill and Les? We all can't go traipsing into their offices all day."

Peter had an answer for that, too. "Use the phone. In fact, we can use the phone even in this office. It'll be more like real diplomacy. We won't be able to tell who's talking to who."

"Whom," Ray corrected.

"Anybody whom's playing," replied Peter.

"No, I was correcting your English. It's whom, not who."

"And I was ignoring you," replied Peter.

"Now I know why we couldn't finish a game in eight hours," Bill added.

Andy thought for a minute while this was going on, but before he could say anything, we all agreed it was worth trying. He agreed to bring in the board the following day for copying, and Peter promised to have his girls make the first set of maps for the first day of play.

We would replay WWI as a board game in the Pentagon!

CHAPTER 27

"Coffee, Long Walks and Colored Pencils"

When Peter brought us the colored game maps, it generated a serious discussion. The first concerned color choice. No one had a problem with Russia being red, but an argument developed over who should be yellow, with the vote split between France and Italy. Italy won, leading to the inevitable jokes about Italy finally winning something. There was also a discussion about how light green and light blue were difficult to discern. Peter agreed to ask the girls to press harder. After a day, we got back together to discuss crayons for coloring. Most of us carried the colored maps around in our pockets or in folders. Both methods resulted in the maps being smudged, so we chipped in and bought the girls colored pencils. It took three days to resolve those problems. No wonder we took so long to make a game move!

Peter had been in the office when I arrived, but other than the normal pleasantries, I had few interactions with him. A polite young man, cheerful and friendly, bubbly some would say. He didn't have a permanent assignment in the office, but moved from desk to desk as needed. Most of the time he spent with Nettie, a matronly civilian whose desk sat in front of the cubicle shared by the Major and the Colonel. Andy said she used to do the things we do, until the work overwhelmed her.

Peter was a curious little guy, best described as impish, a man-child. He exaggerated a resemblance to Alfred E. Neuman by parting his hair in the middle, and cultivated a cow-lick. A perpetual smile never left him, nor did the bounce in his step, a bounce he exaggerated by ending each step on the toe of his trailing foot. Born and raised in a small town in the northern part of Minnesota, he and his family found life in the D.C. area hectic. From the beginning, he sought ways to either shorten his time in the service, or to transfer closer to home, all unsuccessful.

At work, we settled into the Diplomacy game easily. At first, we tried to make all our moves in the first day, but the Major became suspicious of our activity. When Andy reminded us we had a week, the pace became more relaxed. The game became part of the normal eight-to-five routine.

About this time, the 902nd Military Intelligence Group asked me to be its liaison. This was not part of Andy's job, or for that matter, anyone in the office. Mike Allen, the spook I had met after the telephone call from Laos, approached the Colonel for help, and he recommended me. The 902nd MI Group managed the internal security of the Pentagon and the Military District of Washington. An Army group, but most of their guys wore civilian clothes. I asked what they did, and what they wanted me to do.

Allen explained their responsibility for protecting sensitive documents, such as Secret and Top Secret material. As an example, he described a recent exercise in which two of their guys, dressed in civilian clothes, went into an office that had sensitive documents. One would try to distract office workers, while the other would try to "steal" documents marked Secret or Top Secret. Other times, they entered offices at night and looked for classified documents left on desks that belonged in locked file cabinets. They also looked in wastepaper baskets for carbon paper. Typists were supposed to put carbon paper in a "burn bag," along with documents, and dispose of them properly.

The work sounded like an official sanction of college pranks, and I said so. Allen turned serious and told me the 902nd and the Pentagon took their responsibilities seriously. After a recent nighttime raid of offices, they made copies of the documents, redacted the sensitive parts, and taped them to the offending office door. Several officers in these raided offices recognized they had allowed a major breach of Pentagon security and resigned, but not until they had fired several people.

Okay, it was serious shit, not college pranks, but I couldn't imagine myself being trained as a spook.

They needed me for a less glamorous aspect of their job. If they had concerns about security in an office, they would send one of their spooks into the office as a worker, and that's where I came in. Once they'd identified an office they wanted to infiltrate, it would be my job to make an opening there. To accomplish that, I'd make up a phony opening, or if that proved too difficult, make an incumbent disappear for a while.

"You want me to make somebody disappear?" It sounded like James Bond stuff, like they wanted me to kill someone, and scared the shit out of me.

"No, no! All you have to do is make the person disappear for a while."

"Oh, kidnapping rather than murder?"

"No, just reassign them on a temporary duty to a more important job. For instance, if we wanted someone in your office, you'd take someone who's here and assign them to the Joint Chiefs of Staff. Then there'd be an opening in your office, and you could fill it with our guy. If we wanted someone in JCS, you'd assign someone there to go TDY to the White House, and you would fill the JCS slot with our guy. At the end of investigation, everyone would go back to their original job except our target, and he'd go to Leavenworth."

"How often does this happen?"

"About once a month."

"Wow, we have that many spies in the Pentagon?"

"No, it's not just about spies. Most of the stuff we do now is about drugs. If we find someone's using drugs, we look at how they're supporting their habit, see if they're selling, or see if it's interfering with the work they do. There's a lot of dope use all over the Army, and we want to see if there's an organization using the Army as a front. Most times, we find it's one guy. We tell him we're on to him. Tell him to straighten out, and leave him alone."

"All right, something to add to my resumé!"

"No, at least not now. The only one who knows what you're doing is the Colonel. Keep it that way. If we have anything for you, we'll come right to you. It'll look routine."

"What about Andy?" I asked, because I was still in the learning mode, and he often asked what I was working on.

"We looked at Andy for this assignment. He would have been our first choice, but he's getting out soon. You're taking his place. Don't worry, it's under control with the Colonel."

As he left, he told me to write up orders for the Colonel to sign transferring me to the 902nd. He responded to my slack-jaw, air-sucking reaction saying, "Not to worry. It's just a paper exercise. If you do work for the 902nd, you mut be 902nd. You can still keep your IBM card under your blotter."

So, we left it that way. Mike Allen would contact me when and if the 902nd needed me to do something.

As I learned more about the guys around me, I noticed some of them had peculiarities. Across the hall from our office, they had converted an open area to an office by erecting half walls. Two soldiers worked there—a Spec 5 getting out soon, and his replacement, a PFC. This area housed the coffee pot for three offices, so they made coffee each morning, and kept the coffee going all day. The coffee pot was a big, industrial-size percolator with a fifty-cups capacity. In the morning, they'd pour two pounds of ground coffee into the basket, fill it with water, and turn it on. Nobody but lifers, who like to chew their coffee, could drink any after the first dozen cups each morning. Over the course of the day, the brew became increasingly more unrecognizable as coffee, because the electric pot stayed just below a boil all day. As the afternoon wore on, the brew became bitter, strong, and stale. Rather than make a new pot, the coffee guys would add water to the grinds in the basket. The coffee was terrible, but after a few months, I became a convert to this lifer's brew. I always had a warm cup at my desk, which means I had about six to eight cups of this every day, except weekends, when I would get severe headaches from caffeine withdrawal.

The coffee guy's real job was to handle the IBM cards for the three offices nearby. These offices made the assignments for all the Army enlisted men below E-6, for assignments around the world. These cards changed every week. It sounded like a lot of work, but it was only two days of heavy lifting. One day they distributed new cards when they came in, and the following day they destroyed the old cards. This left them with a lot of free time for three days, when all they did was make coffee.

The Spec 5 in charge of these cards had a finance degree from a major business school. During the day, he spent part of his time studying the financial markets and reading books about stock market trends. We referred to him as the "Frito Bandito." He wasn't Hispanic, just Irish with a dark complexion. The name came from his mustache, which made him look like the cartoon character promoting Fritos. His mustache projected beyond the edge of his lips, so it wasn't Army-regulation. He was in a constant battle over it—not so much with the officers and senior enlisted personnel in the nearby offices, because they accepted a lot of our eccentricities as normal. When he learned the Army was considering a rewrite of the regulations addressing haircuts and mustaches, he volunteered to be a model for the unacceptable version.

During his participation in the trial, which lasted thirty days, they gave him a letter, explaining the program and exempting him from the existing regulation. He used the letter for almost a year. When asked if he worried about the consequences of carrying on the ruse, he answered, "What are they going to do, charge me with impersonating a cartoon character?"

Because I was one of the few non-lifers who frequented the coffee pot, we got to talking. He started our conversation by asking me, "How can you drink that shit?" while sipping a can of Coke. He never drank coffee during the day, only Coke.

As time passed, I learned about his interest in the financial markets, and had some meaningful discussions about them, more like lectures; he talked, I listened. He believed in the premise that market behavior was predictable, and if someone collected enough information, they could predict future behavior. At one point, he gave me a book of monographs published by MIT Press entitled, The Random Character of Stock Market Prices, which, despite the title, supported the thesis prices were not random.

The Frito Bandito also had a fascination with the Pentagon. He was an encyclopedia of Pentagon trivia. When he arrived at the Pentagon, he had the normal basic curiosity, but the answers to his questions led to other questions.

One morning, after I had known him for only a short time, he asked, "Why does the Pentagon have so many bathrooms?"

Thinking it was a lead-in to a joke, I responded, "Because everyone here is full of shit?"

The look he gave told me I was wrong. "No. At the time they built it, the country was segregated with bathrooms for whites and blacks, 'colored bathrooms,' they were called. So, they had to build twice as many."

"Didn't know that."

"Now you do."

Following other quizzes, I learned the Pentagon was the largest office building in the world, with twenty-nine acres of office space, equivalent to six-and-a-half million square feet, with ten radial hallways, a five-acre pentagonal courtyard in the center, and the workplace of over thirty thousand people. Other facts he challenged. He understood plans for the building existed before WWII. He believed the plans were not for the home of the Department of War, as it was known then, but for a large military hospital. As evidence for this, he pointed to the ramps between each floor, designed to aid in the transport of patients on stretchers. Of course, the ramps weren't for transporting patients, but to save on steel during the war. Ramps needed less steel than elevators and elevator shafts did.

Another fact he challenged was the seventeen-and-a-half miles of corridors in the Pentagon, but he never made it clear whether he thought the number was too high or too low. He felt it wrong, and became determined to get the correct answer. He set aside time each day to walk and measure the corridors of the Pentagon. On the two days he had IBM card work, he only took short walks. On the other three days, he spent as much time as he wanted. He had a map of the Pentagon, and would color in each of the corridors he walked. He set a goal of walking all the corridors before his time in the Pentagon ended. A quiet man, he became even quieter near the end. I speculated he was running out of time to finish his corridor-mapping mission. I joked with him he should re-enlist to finish, but he didn't respond to any of my taunts.

Then one day, a month before his ETS, he became very animated. During his quest, he had found the true length of the Pentagon corridors—as listed in the book he used as his source—hadn't considered the tunnel running from the Pentagon to a heat plant located outside of the building.

"Once I get into that tunnel and walk the corridor, my work here will be done. I'll have walked all the corridors in the Pentagon."

The end of his quest came a few days later. Returning to the office, his face beamed. On his map of the Pentagon, he drew two parallel lines, from the Pentagon to a box he had drawn outside of the Pentagon. Then, with a flourish, he colored the corridor he had made, put his pencil down, and stared at the paper. He had a peculiar smile, something you'd expect to see when someone has a sense of accomplishment; but underneath that smile, something else percolated. When I visited the coffee pot the next morning, he sat at his desk with his map and the same look on his face. The PFC frantically transferred heavy boxes of IBM cards with his usual sense of purpose, but without the help of the Frito Bandito, who remained staring at his map.

The following Monday, the Frito Bandito didn't show up for work. He didn't report to sick bay, and he didn't call in. Later, when they checked his room in the barracks, they discovered he had cleared all his stuff out. The MPs called his parents in New Jersey, and they either couldn't, or wouldn't, say where he was. The company clerk officially listed him as AWOL, and later, the Army declared him a deserter. All this with less than two weeks until his military obligation was over. We knew he cared little for the Army and its regulations. Less obvious, until the end, was what he perceived as his mission while in the Army. It wasn't to do his two years, or handle the IBM cards, but rather to walk and map the corridors of the Pentagon. When he finished that job to his satisfaction, his job in the Army was finished—not when the Army said so, but when he said so.

The civilians in my office didn't interact very much, keeping to themselves or their work groups. Mrs. Ramirez's people didn't even spend a lot of time talking to each other. The other groups were small, one to three people, which should encourage interaction. As the weeks passed, I noticed this caused a lot of inefficiency. With groups that had similar activities, and sometimes a handover, the work at the point of handover was usually duplicated. Was this an issue of trust, a concern about the others ability to do their job, or merely protecting turf and job? Regardless of the reason, it created inefficiency in the office.

Pam was the exception to these barriers. I noticed this because I always watched Pam, who had taken the place of Julie in my fantasies. I couldn't smell her unless she walked passed my desk, but what filled my eyes more than made up for the scent I lost. Pam worked with the speed-typist and dealt with the soldiers in Vietnam who wanted to re-enlist, and with the soldiers who volunteered for service in Vietnam. I asked Andy why the civilians treated Pam differently.

He smiled. "Pam's not a threat to civilians who've been here a while. She's only here as long as her husband is assigned to the Pentagon. When he goes, she'll go, and everyone knows that. She's not trying to carve out an empire; she's just trying to do a job. Pam's a lot like us, temporary duty. But Pam is drop-dead gorgeous. When the other ladies, and some men, have a problem with a guy who might just be a little argumentative, they ask Pam to do it. Somebody new from down the hall may take issue with Nettie, but when they have to deal with Pam, they melt. Most guys don't want to argue with her. They want to say 'yes, dear.' She is the ultimate unfair advantage."

Louise and Thelma proved to be other exceptions. Louise made the assignments to the Defense Language Institute and the Special Services

Branch. I couldn't figure out what Thelma did, but the two of them enjoyed their jobs, each other, and the young soldiers who brought a spirit to the office while not diminishing the serious nature of the work.

I started paying attention to conversations taking place in the office. Everyone knew everyone else's business. To fit in, I had to listen in. The currency of trade in the office was gossip. You could collect gossip for a while, but then you had to contribute, or they cut you off. The gossip was generally innocent. These were nice people who led ordinary lives with ordinary administrative or clerical jobs. Instead of sheet metal workers or school teachers, they happened to be dealing with the lives of young men fighting an unpopular war. They did their jobs with dedication and reverence. The young soldiers didn't know these folks and didn't feel very lucky or blessed to be in Vietnam. They didn't know the people who sent them there cared about them and worried about them. The soldiers didn't know we were happiest when we brought them home, and tried our best to get them near their loved ones. So what if we shared a little gossip?

As I passed the Sergeant Major's desk one day, I heard him on the phone. The Pentagon is referred to as the Department of Defense, or DOD. I didn't hear but a snatch of the conversation, but it seemed so typical of him, I had to pass it around the office. I considered it a tribute to his wit. Others considered it gossip because it was overheard. What I heard was the Sergeant Major responding with exasperation, "Lieutenant, I can do a lot but not that. I'm DOD, not GOD."

That night, on the ride home, Stoner asked me if I had passed around the story about GOD versus DOD. I was embarrassed to be caught eavesdropping and then repeating what I heard. I responded I was guilty.

"Good. I was afraid you didn't hear me, with all the noise in the office."

"So, you knew?"

"Of course, I knew. Every time you walked to the front of the office, I picked up the phone, ready to use that line. I must have tried four times before you heard it."

"You planted that line on me? You expected me to repeat it? Why?"

"You weren't in the gossip-giving business yet. You had to have something to trade, so the civilians would let you inside. Besides, it was a great line. I thought of it a couple of weeks ago. Been waiting to use it. It's a great line, but it had to be done just right. It had to be me. No one else could pull it off, and it had to be on the phone to someone no one knew."

"You're a crafty old fucker, aren't you?"

"Your response should have been, 'Yes, Sergeant Major, it was a great line. Thank you.' As far as the crafty old fucker, only Mrs. Stoner gets to call me that. You can say I'm a wise Command Sergeant Major, whose years of experience and wisdom you can only hope to mimic as you get older."

Responding quickly, I said, "That's exactly what crafty old fucker means!"

He left it at that.

The modified way of playing the Diplomacy game went well. At first, we decided Andy would adjudicate the moves on Thursday night, but this put too much pressure on Peter's girls to return the new maps on Friday. They would have to color eight maps with seven colors; the extra map to serve as a master for Andy to record the moves. Friday wouldn't work because everyone wanted to get out of the office to start their weekends. Instead, we decided to give Peter's girls until Monday to provide new maps. We had a general idea of what was going on, so it was no big deal. We settled into a routine of discussions over coffee or trips to the men's room. Bill and Les continued to do a lot of negotiating by phone, but occasionally some of us would pay a visit to their offices down the hall. When we needed a longer discussion, lunch was the preferred time.

CHAPTER 28

"Nettie and Peter"

Over time, Nettie's job evolved into routine work, primarily screening individual soldiers who volunteered for Drill Sergeant School or for assignments to AFEES stations. She did a lot of the things we did, but she worked at a much slower speed. Because I only saw the non-routine stuff, Andy and Nettie agreed I should spend a week with her, learning her job, because her work shifted to our desks when she was on vacation, sick, or when she fell behind. Being a long-time government employee, she got a ton of vacation and sick days. The work had to continue. The soldiers in the field waiting for action on their paperwork shouldn't have to wait, just because of an empty chair at Nettie's desk.

Peter worked for Nettie, if you could call it that. She guarded what she did, and never wanted to get ahead. Despite having me and Peter working for her that week, she still only put out ten pieces of correspondence a day. Unlike Andy, with his standardized paragraphs, Nettie wrote every letter from scratch, and prided herself she crafted each letter to differ slightly from the one preceding it and the one that followed.

Nettie occupied the desk of supreme importance, directly in front of the Colonel and the Major. Sitting closest to the SECRET cabinet, her back faced Stoner, with no desks in between. A place I didn't know was considered a place of honor, until she told me so. When I asked Andy why this was such a special designation, he said, "Because Nettie said so."

Nettie looked like everyone's grandma, short and stout with short brown hair going to gray. The flowery dresses she wore didn't seem as businesslike on her as they did on Mrs. Ramirez. A woman who smoked, she had a sexy, husky voice like Susan St. James, that she used on the phone like a school girl. What would the men on the other end of the line do if they found out the sexy voice was in a Sophie Tucker body?

During the week with her, I noticed she, like most of the people in the office, had no personal memorabilia on her desk. At least I thought she had nothing, until I saw a Marine Corps medallion perched on the pedestal of her desk lamp. When I asked about it, she got a little teary. Taking a slow, deep drag of a Chesterfield, she looked off into the distance before saying, "That was the collar brass of my late husband."

Surprised. No one had said she was a widow, or a military wife. "I'm sorry. Didn't know you were married. Did you lose him in Vietnam?"

"Oh, no. We retired before Vietnam got serious."

"We?"

"Yes. I was a Marine. I retired as a Sergeant First Class after twenty-two years. Alf, my husband, also held that rank. You didn't know?"

"No." No longer surprised, I was flabbergasted. I couldn't imagine the lady sitting next to me being a Marine, the fittest of the fit.

She giggled and smiled as she read my look, and said, "I was a lot thinner then. I met my husband on my first tour. We were both stationed over there, at Henderson Hall," she explained, pointing to her right in the general direction of Henderson Hall—the equivalent of the Pentagon and Fort Myer for the Marines. A small military base east of Fort Myer, separated from it by a fence to keep the Army out.

"Alfie was a Lance Corporal, and I was a clerk. We hit it off right away. He was so handsome, and believe it or not, I was quite beautiful. When I joined the Marines, all my friends back home thought I was a lesbian." She giggled again, that throaty, sexy, Susan St. James imitation. "I wasn't. I thought about men all the time. At first, I thought about joining the Army. What could be better than to be a good-looking woman surrounded by a bunch of horny Army guys? Then I saw a Marine parade. The answer was to be a good-looking, horny woman surrounded by a bunch of good-looking, horny Marines. My first Marine took my breath away." She took another drag of her cigarette and exhaled slowly.

Nettie had my full attention, not only for what she was saying, but for the images she was generating in my mind. I wanted to ask if she had pictures of herself as a young Marine, but I was afraid she'd know my motive. I also wanted to ask her how many guys she had before she met her husband, but didn't.

"I'm getting all tingly just thinking about it," she sighed, the exhaled smoke adding to steaminess of her story. "Well, my husband and I, like I said, we hit it off right away. I knew he was the one. He curled my toes." Reacting to my questioning look, she explained, "If your woman hasn't told you that yet, you ain't doing it right. My Alfie did it right. We started shacking-up right away. A little house near the base became available, and we rented it. Before he shipped out for a short tour in Korea, we got married. With his extra Korea pay, we bought the house, three doors down from the main gate at Henderson Hall. I still live there."

With her cigarette finished, she stubbed it out, sat up straight, dusted the ashes from her ample bosom, and got back to shuffling the papers on her desk.

Peter had been standing nearby, listening. "I've been working with you for months," he said to her, "and I never knew any of that. How come you told him, and not me?"

"Two reasons. The first is you are never here long enough for me to tell you anything. You're always hopping around doing whatever. Second, he seemed interested."

"What happened to your husband?"

"He died of cancer eight years after we retired. I started dating Jack when he died. We've been going steady ever since. Jack helps me remember the good times Alfie and I had. Jack also helps me forget how lonely it is without him."

"Well, I'm glad you found someone. Is Jack a Marine?"

"Jack is everywhere and is everybody. Jack was a good friend of Alfie and me. I wake up with Jack every morning. Have breakfast with Jack. Sneak a little visit with Jack at lunch. At the end of the day, I go home to dinner with Jack, and I fall asleep with Jack…Jack is Jack Daniels." Looking off to a memory I couldn't see, she paused, then continued: "I'd trade all my time with Jack for the rest of my life, for one minute with Alfie."

Speechless, I just stared at her.

"I'm a functional boozer," she went on. "When I cease to be

functional, I'll quit, and then I'll die, but until then, I'll help these boys who want to be drill sergeants and recruiters get to live their dream."

"Why don't you work for the Marines? Henderson Hall is closer, and you have a history."

"Too many memories. I tried it there for a while, but everyplace I went, I remembered Alfie." She giggled again. "Hell, if the truth be known, during our marriage, Alfie probably banged me every place in Henderson Hall you could lie down, stand up, or bend over."

I thought of the Frito Bandito with his map of the Pentagon corridors, and tried not to imagine Nettie and Alfie with their map of Henderson Hall.

Nettie passed me off to Peter for the rest of the day, suggesting he show me what he was supposed to do, not what he did. We moved to my desk.

"So, Johnny. What do you want to know?"

"Johnny would like you to call him Jonathan, and then tell him what Nettie and Peter do."

"Sure. We do some of the stuff you and Andy do, only we do less, and it takes us more time, and we're never finished."

"And why is that?"

"Because you and Andy do most of what we do, you do it better, and you do it faster."

"And why is that?"

"Because Nettie is a civilian who really doesn't care, and I am a soldier who doesn't want to be a soldier, at least not here. If I have to be a soldier, I would rather be a soldier in Minnesota."

Peter told how he got drafted right after graduating from college in northern Minnesota, in a town known by the network news as being the coldest spot in the continental United States, at least five times a winter. He was married and had two daughters, but got drafted as soon as his student deferment expired. Like most of the guys in the office, he came from the clerk school at Fort Dix. Delighted at first, his family joined him, but the cost of living in the area soon became a burden. Both he and his wife worked a part-time job, but they still couldn't make ends meet. They found it difficult to adjust from a small rural town in Minnesota to the rapid pace of the nation's capital.

"In seventeen days, I'll have less than one year to go in the Army." He said this like it was one of the most important dates in American history. Sure, a lot of guys are happy when they turn the page on their ETS calendar and it reads less than a year, but to Peter, it seemed religious.

"Why is that so important?"

"Because, Johnny—"

"Jonathan," I corrected.

"Because then I'll no longer be eligible for assignment to Vietnam. No permanent change of station, except from Vietnam, at Army expense, if you are less than one year from ETS. Andy told me that. So, no matter what I do, they cannot send me to Vietnam."

"No matter what you do, sounds like you have something planned."

"Yes, I plan to ask for a Permissive Reassignment to the ROTC unit at Bemidji State University in Bemidji, Minnesota. That's the closest place to home. We can live with my in-laws, my wife can get her old job back, and the kids will be able attend our old school."

"What if they don't have a slot for you at Hot Diggity U.?"

"Bemidji State University," he corrected. "They do have a slot. They've had an opening for three months. The IBM card is under my blotter."

Seems Peter had a plan. We chatted for a while longer, and he left, off to do whatever he did. When he left, he said goodbye with his characteristic adieu, "See you around the campus." This was sometimes preceded with a casual, "O-KEE-DO-KEE", or if in a hurry, "O-KEE-DOKE."

"What a goofy walk! I never noticed it before," I said as I turned my chair back square with my desk to face Andy.

"Don't even go there," he responded.

"Go where?"

"The walk. Right after Peter got here, after he settled down, he started doing goofy things. We had a guy, Al, who was connected on the Hill. His uncle was a senator. Angelo was short, almost too short to be in the Army, but I guess with politics in the family, he wanted Vietnam service time on his resumé. So, he enlisted, then managed to get stationed here. Everyone in the office heard about his uncle, but it didn't make any difference. He was smart, worked hard, and never took advantage of his connections. Angelo had a little bouncy walk, and Peter copied it. Also, when he walked, he'd squat down to the same height as Angelo. The Major called him on it—no walking around short. Peter noticed I have a little bouncy walk too, so when the Major told him to knock it off, he said he was walking like his hero, Andy. The

Major thought that was funnier than walking like Angelo, and so it was okay. Things were weird here for about two weeks when he started it. It looked unnatural and exaggerated, but then we got used to it."

I told him about my two hours with Nettie and Peter, about Nettie being a Marine, Peter wanting out of the Army, or at least out of Washington, and us doing their job. Andy told me now I knew everything the rest of the office knew, but I still had to work with them for the rest of the week, so my official training record would show familiarity with what they did.

"Are you kidding me?"

"No, the Major has to keep a CYA log of—"

"No, I get it, but that's not it. Everyone knows Peter wants out of here, and wants to go to some place in Michigan?"

"Minnesota."

"Whatever! Everybody knows?"

"Yup. Everybody knows he's got the requisition card under his blotter. Everybody knows he's up to something, starting in seventeen days."

"Even the Colonel?"

Andy looked tired of answering my questions, answers I could get from anyone in the office. "Not sure about the Colonel, but the Major does. He's looking forward to what Peter will dream up in the way of stunts. He thinks Peter's a funny guy, and smart, too. The Major thinks it'll be fun to watch and see what he comes up with."

"Will Peter get his reassignment to Minnesota?"

"Not at first. It would set a precedent. If Peter gets it without a struggle, what's to keep anyone from pulling the same stunt? Peter will get it just before he does something stupid that could have a negative effect on the office or the rest of his life. Remember, we're the good guys here in this office. We're here to help the soldiers, not let them screw up, even if them is us."

With that, Andy turned back to his work, and I went back to pretending I was working with Nettie and Peter. To make it seem real, and to make progress on my work load, I took my work over by her desk and did it there. When I had to type or use the phone, I came back to my desk.

Andy had an idea. "When we were talking about Peter getting away with it, but putting him through some hard times until then, it occurred to me your recruiter should be put through a rough spell. Let's call him."

The recruiter answered on the second ring. "Sergeant, this is PFC Jonathan West. Do you remember me?"

"West, old buddy. How you doing? Are you still in the Pentagon?"

"Correctomondo, Sergeant. Still at the Pentagon, and still in EPCMR-GS. I am responsible for worldwide assignments for all seven, eight, nine, and zero MOSs for all E-6 and below soldiers." I paused, letting the full impact of what I had just said sink in. "Let me tell you just how smart this ole college boy is, Sarge. Right now, you just figured out I'm responsible for your worldwide assignment. Am I correct, Sergeant?"

"Hey buddy—"

"I'll take that as a Yes, because we talked about that the first time I called you. The other thing I wanted to tell you is, the Army thinks I got a raw deal on the Special Services hockey thing you allowed me to sign up for. They called it a fraudulent enlistment, or something like that, and you're the fraudulator. Anyway, they said I don't have to stay in the Army for three years; only have to do two. Ain't that great? And, the best part is I can stay right here in good ole EPCMR-GS, taking care of my buddies, like you. Nice talking to you, Sergeant."

Andy was all smiles, and knew I just wanted Jenkins to squirm a little—well, a lot. I didn't want any harm to come to him. I wouldn't send him to Vietnam for spite. It felt good, being safe, and in a job where I felt I did some good. And now, emotionally, I had had a little taste of payback on a man who used his position to take advantage of me at a scary point in my life. No doubt he had taken advantage of others, too, adding to my feeling of payback.

Andy increased my responsibilities as soon as I finished my week with Nettie. Each day, the phone was ringing at eight o'clock, with a call from the main Personnel Office at USARV HQ in Vietnam. USARV was the United States Army, Vietnam, one of two major Army personnel centers in Vietnam, the other being MACV, Military Assistance Command, Vietnam. One job they had was to make sure every soldier rotating back to the States had a place to go. Ordinarily, they did, but occasionally, a soldier fell through the cracks with the paperwork, lost somewhere along the long chain of command. It was a long way from USARV HQ to the boonies where a lot of these guys were stationed. Sometimes, guys showed up ready to leave, and had no place to go. These guys couldn't get on an airplane back to the States without orders to some place in the States, and remained stuck in the departure lounge. So, Andy set up this system where the Personnel guy in USARV would call with the names, MOS, and a preferred assignment in the States for those waiting. With Vietnam holding on the phone, we'd go to the cabinet and pull assignments. Then we'd get back on the phone with USARV, give them the assignment. They'd cut the orders right there, and the guy would get on the plane.

At first, only one or two guys a day needed assignments, and with lots of assignments in the States, the system worked well. With Nixon withdrawing troops from Vietnam, the number of guys without assignments increased, and it became tougher to give the quick turnaround. This was for two reasons. The phone system the Army used was called AUTOVON, which stood for Automatic Voice Network, and had a listening device on it. If there was no talking after three minutes, it cancelled the connection. Being a super system, it was busy all the time, especially after eight, when the Pentagon woke up. To fool the system and avoid getting cut off while we looked for assignments, we told the guy in Vietnam to put the phone next to the radio. With the increased number of guys needing assignments was the lack of assignments in the States. A lot of guys were coming back and filling up the Army bases.

The Army bases on the East Coast were the most popular. Except for Fort Dix in New Jersey and Fort Devens in Massachusetts, we didn't have a lot of places to send guys. At first, we'd send these guys to the reception center either in Oakland, California, or Fort Lewis in Washington, hoping something would open while they were on leave. After a while, we said the hell with it, and made up assignments. If a guy said he wanted to go to Fort Devens, we made up a control number and sent him there. Any place these guys wanted to go, they got. After all, they had just spent a year in the jungles of Vietnam. They had earned anything they wanted.

This wasn't without consequences. Each of these telephone

assignments we made had to have an authorization line, and we couldn't make that up. These authorization lines were either, "Per telecon SP5 Henderson," or "Per telecom PFC West." It didn't take long before the telephone rang with calls from the Personnel offices of Fort Devens and Fort Dix. They were pretty much the same: "West, how come you sent me so many clerks (or supply specialists, or cooks)? I checked my records and didn't have any requisitions. Shit, I have soldiers hanging around doing nothing. I don't have barracks space for them. I have to send them home to live with their families. They come in for formation in the morning, have chow, police the area for cigarette butts, and then I send them home. What gives?"

Our standard answer would always be, "Can you read the OPO Control Number from their orders?" Of course, if the OPO Control Number had a Z in it; we had made it up, and it wasn't against a real assignment. These Z codes were supposed to be used for extraordinary circumstances, like a Red Cross case. Only Andy and I had access to the Z codes. We would always respond it was a valid OPO Control Number, leaving the caller muttering to himself. Andy and I enjoyed these calls, knowing soldiers were sleeping in their own beds each night and eating Mom's dinner. I hope the breakfasts they got were as good as the ones at Fort Myer South Post.

Every once in a while, the calls weren't friendly. We referred to these as the "dick-banging calls, usually occurring after one of the friendly calls. If a Sergeant called the first time, the second call would come from an Officer, often a Second Lieutenant. He'd try and pull rank announcing he was a Lieutenant. This was the tip-off that he was at the bottom of the pecking order, probably an ROTC graduate who couldn't go to Vietnam, so he got stuck with a boring state-side assignment. My Basic Training Company Commander found out he was allergic to bee stings after he got his commission.

The conversation was a welcome distraction. Andy introduced me to the sport. I heard him say, "Yes, Lieutenant I made the assignment. It has an OPO control number so it's valid. Yes, Sir. If you want to speak with my boss, I'd be happy to transfer you."

Andy put the Lieutenant on hold and told me to join him at the Major's desk. "Sir, we have a dick-banger on the phone. Wants to talk to my boss." Andy indicated I should stay as he returned to his desk.

"Major Jones here Lieutenant. How can I help you?"

The Major listened for a minute, then interrupted. "The assignment is valid Lieutenant." A pause, then he continued, "Yes, I'd be happy to explain that to your boss, the Lieutenant Colonel,' with emphasis on the Lieutenant Colonel. "I have to tell you, I'm going to tell him the same thing I told you. If he wants to talk to my boss, he should know my boss is a one star. We can do this a couple of times, but the bottom line is my boss is always going to bigger than his boss." A pause again, 'You're very welcome Lieutenant."

When he hung up, he smiled, "That was mild. They don't usually give up that easy. It's more fun when the call is from a senior enlisted man because then they get the Sergeant Major rather than me. The Sergeant Major can say 'my dick is bigger than yours, so you're going to lose," but officers can't say that."

There was always an afternoon call from MACV HQ, identical in purpose to the morning USARV call. Andy and I settled into a routine where I handled the morning call, and he took the afternoon call. We had confidence in the guys on the other end who we dealt with.

This system had the potential for major abuse. If the guys in Vietnam Personnel were larcenous, it would have been easy for them to sell assignments on their end. We made it clear to them their job was to make sure no one over there was screwing with us. If we found out, we'd make sure they never got out of Vietnam, and if they did, they'd be assigned to Fort Polk, Louisiana.

The job was developing depth. There were lots of different things to do, and many people to deal with, but most of all, there were problems to solve every day. When solved, it meant someone was happy, and that made me happy.

Diplomacy remained part of our daily routine, and we fit it in to the normal course of activity. It was exciting, because the game was progressing in a way different from the version recorded by history. This was probably because each of us wanted it to be different. Not that any of us wanted Germany to win, but we wanted Germany to lose without the intervention of the U.S. government.

But history is not to be denied. Our game of Diplomacy, our re-creation of the events of 1914-1917, could not continue without the intervention of the U.S. government. And, like history, when the U.S. government got involved, it became a World War.

CHAPTER 29

"Diplomacy"

Friday, and Andy had taken another three-day pass, leaving me as the "go to" guy. Now, I was full of self-confidence and welcomed the opportunity to put my stamp on the job, to prove myself worthy of taking over when he left. Andy asked me to keep an eye on Peter. Today was the anniversary day he had been waiting for; as of today, he only had 364 days left in the Army.

When Peter came in this morning, the Major spoke as he passed his desk. "One more weekend, Peter, and you'll be under a year."

Peter stopped, faced the Major and responded, "Beg to differ, sir, but today is day 364."

"Maybe, if you don't count leap year," responded the Major.

Peter's shoulders sank, as he shuffled to his chair near Nettie's desk and sat.

Nettie giggled, and said, "The Major's fuckin' with you. This isn't a leap year."

Peter looked at the flip calendar on his desk, narrowed his eyes as he tried to remember something, then beamed as he looked at the Major.

"Sorry, but if this year was a leap year, I'd be right," the Major said with a smile.

I didn't have to wait long for something. It wasn't part of Peter's plan, and it surprised him as much as anyone. A surprise that involved Peter, me, the Major, the Colonel, a General, and the 902nd MI Group.

I was at my desk, chewing on my first cup of coffee from across the hall, when the Colonel called me to his desk. Peter stood there, flanked by two of the guys, one being Mike Allen from the 902nd MI Group.

"Is there a problem, sir?"

"Don't know yet, West, but you're 902nd liaison for us. If they have a problem, I have a problem, and that means you get my problem. Come along."

The spooks led us down the hallway to the Conference Room. A one-star General sat at the head of the table. The further I got into the room, the more people I saw; a few Majors I didn't recognize, probably aides to the General. The closer I got, I could see one Major had the JAG Corp brass on his uniform, the other guy had the Infantry idiot sticks. The Judge Advocate General representative quickened my heartbeat. I wasn't worried about myself, but because the two 902nd guys flanked Peter, I thought he had already gone over the edge. If he had, I had let the Major and Andy down. With Andy absent, I wasn't sure I was up to whatever would happen next. What happened to the confidence I had ten minutes ago?

A civilian woman sat at the table. Frail, she was older, about sixty, sitting with hunched shoulders, hands on her lap, looking scared and nervous. The scared look made her appear frailer, or her frailty made her look more scared. She had short, dark brown hair, not natural, but something from a Lady Clairol bottle. When she started touching her hair up twenty years ago, it might have looked okay, perhaps even natural, but now the color screamed "I'm dyed" on her sixty-year-old head. I didn't know her, but I'd heard of her. Reportedly the longest-serving employee in the Pentagon, some people said she was here before they broke ground, and they built the building around her. Even though the room was warm, she shivered, her hands moving up to clutch the sweater she wore.

Deep in the corner to my right, I almost missed Bill Eastman, who stood with another of the 902nd guys. Was Eastman 902nd for his office? This thing must be big to have so much brass and so many 902nd guys.

The little old woman screeched, "That's him. That's the one." She pointed at the door, rising a little in her seat, but not attempting to stand, more of an exercise in being seen and heard. At the door, there was only Peter, followed by two 902nd escorts.

Peter smiled at her and gave a little wave and his trademark, "Hi di ho, Mrs. Dropov."

This put her into a rage. Now she stood, jaw set and as ferocious as her frail body would allow. "Don't you 'hi di ho' me, you pinko, commie bastard! You're a fool if you think you could get away with this. When my husband and I came to this country, it was to get away from scum like you."

The General broke in, touching her arm to calm her. The gesture had the desired effect. His sympathy either soothed her or communicated to her he was on her side. She sat back.

The General spoke to the room. "Mrs. Dropov has raised a concern. She has observed some activity over the past several weeks that requires us to examine her concerns. All of you are here to listen to what she's witnessed. All of you are here for a reason, either as a principal in what Mrs. Dropov has seen and heard, or, if we do not refute her allegations, then you will be involved in one of the most serious investigations ever to take place within this building. Others may become involved, but this is where it will start. Now, I will ask Mrs. Dropov to tell her story, to tell what she has seen and heard over the past several weeks. I'm going to ask that you extend her the courtesy of not interrupting her or asking questions until she's finished. Both the JAG Office and the 902nd have requested we record this meeting." With this short introduction, he turned the meeting over to the JAG Major.

The JAG Major asked everyone in the room to identify themselves and sign a paper he passed around. Sign our name, print it, provide our office designation and a work phone number. This sounded very serious. Andy, I wish you were here today. I feel like I'm in a courtroom.

Mrs. Dropov began by explaining how she and her husband had fled Poland in 1939, just before the Germans had invaded the country. First fleeing east into Russia, and then north to Finland, where they escaped to the United States. She described how she and her husband worked hard to learn English. She studied at a secretarial school and earned a position in the War Department, later the Pentagon. Mr. Dropov was restricted to menial jobs, because he had no ear for English, more like no tongue for English. While he could understand English, he spoke it poorly with a strong accent. A smart man, but his accent limited his opportunities, until he took a job with a limousine company.

Peter broke the bored silence with laughter. All eyes turned to him. Grateful for the interruption, as I'm sure others were, but decorum required us to view Peter with disdain for violating the General's request for no interruptions. "I'm sorry," Peter said, "But, that's funny." Blank eyes stared back at him. "Mr. Dropov works for a limousine company? Get it? If he had a partner named Alexandre Pikop, they could call their company 'Pikop and Dropov Limousine.' Get it? Pick up and drop off limousine?"

Everyone got it, and most of us thought it was funny, judged by the number of guys, enlisted and officers, who coughed and sneezed to mask their amusement. The General asked Mrs. Dropov to continue and to focus on the things she had seen in the last several weeks that gave her concern.

Continuing, she described how her desk was across a narrow aisle from Eastman, whom she described as a loner, someone who did not interact with the others in the office. Eastman stood up straighter, surprised he was the target of her concern, and cringed at the word loner, a word associated with psychopaths and serial killers by the press. Being the only enlisted man in the Congressional Liaison Office, his job as the office typist and go-fer for the officers would make him appear to be a loner.

Eastman was anything but a loner. He was a tall, well-educated, good-looking guy. Physically, he could have been Les's brother, just without the accent and the fixation on women. Eastman fixated on his career and money. Before being drafted, he worked for a large foreign insurance company in New York City, where he was on a fast track. When his draft notice came through, his company tried everything they could to get him off. They failed, but to show Eastman the high esteem they held him in, they promised him his old job when he got out. Eastman didn't think this was such a big deal, in fact, he argued there might be a law requiring them to do so. If no law, then common decency would dictate a foreign company would keep a job open for an American going off to war. Besides, he argued, he might be more valuable to someone else when he finished his military obligation. When they offered to continue to pay his salary to his wife while he was in the Army, he stopped arguing. After he got assigned to the Pentagon and landed the job in Congressional Liaison, he felt it was time to renegotiate his contract. After all, he reasoned, an American company might find him even more attractive if he had contacts in the House and Senate.

His bosses had no idea what contacts he had with Congress, but the title of his job was certainly impressive. When they asked him to describe the job, he embellished it, never straying far from the truth, but not defining his role versus the role of the office. The company encouraged him to get letters of gratitude from the people he helped, be they in the White House, Congress, or among business leaders whom he helped. Letters of gratitude are the stock in trade of Congress, and he collected a lot. Most weekends, he drove his E-Type Jaguar to his home in New Jersey, and spent the weekends socializing with his insurance friends at the country club. No, he was not a loner; he was a very social person.

Mrs. Dropov described recent changes in his behavior, notably, more time on the phone having secretive conversations. The most damning charges she made were accompanied by an increased agitation in both her voice and body movements.

"Then, about three weeks ago, that little man with the glasses"— she pointed at Peter—"began to visit him regularly. They talked in whispers, and sometimes had their heads huddled together. I thought at first they might be...you know. But then yesterday, I overheard what they were talking about."

Sitting up straighter in her chair, she paused for effect, looking around the room before continuing. "The little man said he had just talked to Germany, and Germany agreed to support the attack on Russia. Then he said France, England, and Italy had also agreed. He said it was something that should have been done during World War I, and it would have changed the course of history."

All eyes in the room focused on her. When she stopped talking, the room was silent except for the whoosh of air coming in from the vents in the ceiling. The room was so quiet, I could hear footsteps in the hallway outside of the closed room, on the carpet!

Then all hell broke loose. Eastman stepped forward from his place against the back wall.

"It's a game!" he shouted, breaking the silence.

Peter stepped forward, a serious grimace on his face, hiding the impish twinkle in his eyes. "That's the problem with everything today. Everyone thinks everything is a game. This is not a game. It's a World War, and it should be treated seriously. Listen to the people on the streets, they don't look at Vietnam as a game. They see the horror. War is serious business."

"It's a fucking game," Eastman said as he stepped forward to confront Peter.

If all hell broke loose before, purgatory got demoted and joined the fight. Everyone started talking, all with waving arms and desperate expressions. The officers' faces all registered career-ending disbelief that something like this could happen on their watch. The 902nd guys didn't know what to do, so they stepped forward and grabbed the arms of Eastman and Peter, standing, waiting for instructions.

The Major stood on my right. Leaning in, he asked, "Do you know anything about this? Is this the start of Peter's shit-storm to get out of here?"

"Yes, I know about it, and I don't know if he planned this."

The Major stepped forward and held up his arms asking for quiet. He was the most junior officer in the room, and he was asking his superiors to shut up. Brave or foolhardy, I couldn't guess. What could he hope to accomplish? On top of everything else, he was making himself a target. Wrong! The target was on me.

"Most of you know Specialist West. He just told me he might have something to add."

All eyes shifted to me. The General, one of the Army's first and only black generals, stared at me. He didn't speak, but his eyes and body language were saying, "Well?"

Taking a step forward, I spoke.

"Sirs. This is a terrible misunderstanding. Eastman is correct, this is a game." Peter stepped forward, presumably to start another rant, but the looks from me and the Major pushed him back. The two 902nd guys holding his arms had something to do with it as well. "They call the game 'Diplomacy,' a re-creation of World War I. There are seven players representing seven countries, including England, France, Germany, and Russia." I said this looking around the room, pausing at Peter and Eastman. Relief settled in as those officers seated at the table fell back into their chairs, content their careers would not end by allowing a war with Russia to start under their command. "We started playing the game off site, but it took a long time. We tried it several times, but never completed a game."

"You're playing a game, based on World War I, here, inside the Pentagon, during working hours?" asked the General.

"Yes sir, but to put it in perspective, most of the time, we play the game on coffee breaks or during lunch. It's better that way—no one gets to see who you're talking to."

"Except for Peter and Eastman?"

"Yes, sir. Unfortunately. We probably should have had stricter rules."

The General asked, "How long did you play?"

"We've been playing for about six weeks. The first time, we took about four hours and only made one move. The second time we played for over eight hours, and still only made two moves." Then I explained how we moved the game to the office and made one move a week."

"Where are the others—let's see, Peter and Eastman are two; there are five more players?" The General asked.

"Les is in White House Liaison, and the rest of the guys are in General Support." At this, the Major and the Colonel became alert. They were the only ones who had the look of concern for their careers now; the others relaxed. Answering his question, "I'm a player, as are Andy, Mullaney, and Ray. I'm one of the people Mrs. Dropov heard Eastman talking to on the phone."

"Is that why you guys all hang around late on Thursdays?" asked the Major.

"Yes, sir. That's when Andy arbitrates the moves."

The General leaned forward in his chair, elbows now supporting his hands which formed a steeple. "And you've only been able to make a couple of moves in all this time?"

"Yes, sir, one move a week. The game is complicated, much more than we expected. We figure it'll take us about two months to get to a point where we can declare a winner."

The General took a deep breath, then said, "Mrs. Dropov, I want to thank you for all the service you've given your adopted country, working here in the Pentagon. I also want to thank you for your diligence and dedication in bringing this to our attention. While not a National Security threat, your quick action would have averted a very serious situation, had it been real. I wish everyone was as dedicated as you are. I'm going to excuse you now, while we decide what to do next. Thank you again."

Mrs. Dropov got up and left, but not before glaring at both Peter and Eastman. I got a half glare for proving her wrong.

With her gone, the General asked for the door to be closed, and pointed several of us to empty chairs around the table. His eyes made the same furtive movement around the room as Mrs. Dropov, stopping at the same faces, omitting only the glare.

"I am of two minds here. While I'm glad we won't be responsible for World War III, I am disappointed that soldiers, who I thought were serious and genuinely making a positive contribution to what we're doing, are playing a game during working hours. The game will stop at once. I'm equally disappointed with my senior staff for allowing this to happen, only to be discovered by a civilian clerk, an elderly civilian clerk."

The Major leaned forward in the chair he had just taken. "Sir, I can assure you neither I nor any of the senior staff in our office were aware of this. With most of the game players in our office, we would have seen any unusual activity and stopped it. The fact that only one civilian, with an active imagination and a nosey disposition, caught on, shows the men were aware this shouldn't have been done in the office. As Specialist West pointed out, most of the negotiation part of the game took place during trips to get coffee or during lunch."

"Major, I tend to agree with you. The men involved were careful about their conduct. Mrs. Dropov does have a reputation as being a bit of a busybody with everything other than the work on her desk. It's understandable from her background. She comes from a country where she learned to fear strangers and fear authority, all leading to an increased sense of self-preservation. I doubt she'll be making an issue of this beyond what she's already done. To go public, would be admitting she's a busybody, but more important, she would have to admit she was wrong, something she wouldn't do. I consider this matter closed, and I ask the 902nd to also close the book on this. Disciplinary actions will be left to the Branch Chiefs; but let me add I would be disappointed if that disciplinary action found its way into anyone's permanent file. This meeting is over."

On the way back to the office, the Major came alongside of me, and then slowed down, a signal I should slow down as well.

"Today is Peter's anniversary countdown. Was this a set-up by him?" he asked.

I looked at him and shook my head. "Peter has a plan, but I don't think this was it. That lady coming forward was just a piece of luck he tried to take advantage of. He backed off as soon as I mentioned the other guys involved. I don't think he thought about that when he saw his opening. If he pushed it, it might have taken all of us down, including you, the Colonel, and the Sergeant Major. Peter wants to go home, but not at our expense."

"That's what I think too. He is a clever little man. We're going to let him go at the beginning of August—at least, that's when we'll tell him. He'll be able to send his family back and get them settled in before school starts. Then he'll be able to follow them at the end of the month. Does he still have that ROTC requisition under his blotter?"

"Thank you, sir. I'm sure he'll appreciate that. I'm sure he has the most current requisition squirreled away someplace."

"Don't tell him any of this, but let him know, we're watching him closely. We expect him to pull some stunts." The Major smiled, and with that smooth southern accent, he continued, "We're actually looking forward to see what he has up his sleeve; but if he screws up and embarrasses us, all bets are off. He can still send his family home, but we'll keep his ass here in Washington until his ETS."

"I'll keep an eye out for him, sir."

When we got back to the office, the Sergeant Major came around to all of us, one at a time, and told us the Colonel wanted the enlisted men, no exceptions, to stick around after work for a meeting. We gathered around the Colonel's desk at 1630, but had to wait. A few of the civilians had learned of the meeting and were trying to listen in. The Colonel told them the meeting wouldn't start until all the civilians left. They left. He probably would have asked to have the doors closed if we had any. In the absence of doors, he asked the Sergeant Major to stand at one doorway, and the Major to stand at the other, this in an attempt to keep unwanted ears from listening.

It was the first time since I had joined the office it had taken on an air even remotely resembling a military organization. All of us sensed it, standing in a line in front of the Colonel's desk. As the minutes passed, we stood straighter, our shoulders rolled back, our hands stopped moving and found the seam on the sides of our trousers. No one would mistake us for the Guards at the Tomb of the Unknown Soldier, but we were more soldiers now than we had been since any of us had gotten off the bus in Washington.

The Colonel sensed the change and moved from one end of the line of soldiers to the other, recalling, perhaps, times in the past when he had commanded real soldiers. The moment was broken for me when my eyes met those of the Major, who winked, then gave briefest of smiles, only to return to the stern face.

"I'm assuming you understand why we're here tonight, so I won't waste our time by retelling the circumstances. However much I admire the camaraderie of soldiers who both work and play together, and the spirit such a close relationship brings to a unit, I cannot condone the embarrassment your game has brought to the office, to me, to the Major, and to the Sergeant Major." At this, the Major looked to the floor and the Sergeant Major looked to the ceiling; the Major's body language said, "Don't make too much of this," and Stoner's body said, "Who gives a shit about this? I don't. I'm a Command Sergeant Major, the top of the food chain in this Army, and I've already got my retirement papers in."

The Colonel continued walking and talking. "The General has deferred disciplinary action to me. I have discussed this with the Major and the Sergeant Major, and we agree military discipline has become lax in this office, perhaps leading to an environment that encourages games-playing. We have agreed we will enforce a stricter adherence to military discipline, effective immediately. You have been lax in your appearance. You will stand at attention when I speak to you."

We tried, and it was close enough to satisfy him. It was close enough that the Major smiled like a dad smiles at his eight-year-old son who just struck out for the second time in an inning of a t-ball game. It wasn't close enough for Stoner, who again searched the ceiling tiles for comfort.

"You will also shine your shoes and clean your brass, tonight, and you will show up tomorrow, on time, with regulation haircuts. Do I make myself clear?"

Peter cleared his throat as if to speak, but I stepped forward with a quick, "Understood, sir. No questions," and dared Peter with my eyes to speak. He didn't.

"Dismissed."

The walk to the car with Stoner was quiet. We saluted all the officers we passed, including the Lieutenants and Captains.

Once in the car, Stoner wanted to talk.

"You're all good boys. I'd be proud to have any of you as my son, except Teddy, 'cuz he's stupid and an asshole. He's so dumb, I wouldn't send him to the beach to get sand. Maybe Peter, 'cuz he's too much like my son when he was in high school. That kid drove me crazy. It doesn't bother me what you did. This old biddy overheard part of what you were talking about and put her imagination to work overtime. You boys had the game going on in the office for weeks, and no one was the wiser. Even me, and I know everything. Your work has always been great, and it didn't suffer because of your game."

"Thanks," I said. Stoner looked at me with a look that said this wasn't a conversation, it was his monologue.

"What the Colonel said about embarrassment isn't true. I wasn't embarrassed. I'm at the end of my career. This is a cushy assignment so I can max out on retirement benefits. The Major doesn't care, either. This assignment is a star on his record, and I really mean 'star.' He's young and bright, and he knows how to treat his men as men, not just personnel. He's on a fast track, and the track takes him through the Pentagon, just like I did ten years ago. The Major will be back here someday with a star or two on his collar. The Colonel knows he's not going to get a star. He'll rotate out of here and get his eagle, but that will be it. He sees you guys and knows you'll all go back to being civilians, and be very successful. Some of you might even make a significant contribution. Andy will, and you"—he looked at me—"have a chance, too. Everyone except Teddy, who's an asshole."

Then there was silence. Two traffic lights later, he said, "Now you can talk."

Ray started to say something, but he was using his pompous voice. Stoner interrupted him. "You might make something of yourself, too, if you lose that tone that makes you sound like you think you're so fucking superior." Ray didn't finish.

"Oh, one other thing. The General asked the Major to get the name of the game from you— that diplomacy thing. He also needs a quick rundown on the rules, and why you took so long to play. Seems he thinks his friends might enjoy playing. I'd discuss it with the Major tomorrow, ask if he wants to be your errand boy and take the chance he has to say 'I don't know' when the General asks him a question, or whether you teach the General yourself."

"I'll take care of it in the morning, Sergeant Major. Begging your pardon, but there are no barber shops open after six p.m. We'll be able to shine our shoes and do our brass, but no one will have a fresh haircut. Will the Colonel be pissed?"

"I'll take care of the Colonel, and I'll make arrangements for you to have haircuts tomorrow in the Pentagon barbershop." Getting an appointment there on short notice was tough. "Maybe I'll ask Rocco to give you all haircuts like mine. It'll be quick, just clippers," he said as he ran his hand over the short stubble on his head. He laughed.

The next morning, Stoner made our appointments, two at a time, to get haircuts. Rocco told each of us we were to get a Stoner special, then laughed. We had all come in with our shoes shined and our brass cleaned. When I spoke with the Major, he agreed it would be better if I talked with the General. Once again, life in 1D726 was back to abnormal.

CHAPTER 30

"Peter Starts Writing His Ticket"

Andy laughed when I told him about the Diplomacy game scandal. After discussing what the Major told me about Peter, we agreed we would do our best to keep him out of trouble. The primary goal being to keep him from doing something that would put the rest of us in jeopardy. If necessary, we'd tell Peter he'd be in Minnesota by September if it meant protecting him or us from harm.

As we finished discussing Peter, the Colonel called us forward for inspection. All shoes had an acceptable glow except Peter's, which he looked at more closely. "Peter, you've put some effort into your shoes. While they have a nice shine, they don't look black."

"Thank you, sir. I only had oxblood shoe polish, but I think they came out pretty good."

The Colonel found everyone's brass acceptable, but again, Peter's deserved a closer scrutiny. "The collar brass is on backward and upside down." The Colonel said this looking alternately between Peter's collar brass and his face.

"Sorry, sir. I'm used to putting my brass on in front of the mirror, but last night, after I Brassoed them and ironed my uniform, I put them on. Guess I thought I was still looking in the mirror."

"And upside down?"

"Can't explain that, sir."

The Colonel let us get back to work, but wouldn't let Peter go until he straightened out his brass. The Colonel told him to get a can of black polish. As he walked back to his desk, Peter leaned toward me and said, "I tried three different shoe polishes last night; first I tried cordovan, then brown, and then the oxblood. Cordovan and brown look okay on black, but this oxblood stands out."

"You have all those shoe polishes at home, but you don't have black."

"Oh, I have black," Peter said with a twinkle in his eye, and that's how it began. He was starting to write his ticket to Minnesota. "Thought about letting the girls color my brass with nail polish, but saved that for another time."

When I told Andy Peter had started his race to Minnesota, he shrugged and said, "Buckle up. Normally I'd say it's going to be a bumpy ride, but with Peter, it'll be the bumper cars."

The guys in the office had always been careless about using military time. There wasn't much call for it in our jobs, and the civilians always used a.m. and p.m. designations. The only time we were exposed to it was when we got a notice about something happening at Fort Myer, and then it was fun to watch college grads use their fingers to figure out what 1600 hours meant. Peter decided it was time we used military time, or more precisely, his version of military time.

Military time is not sophisticated. It uses a twenty-four-hour clock, rather than two twelve-hour clocks. The standard way of telling time is divide into a.m. and p.m., twelve hours each. Military time uses twenty-four consecutive hours, with one p.m. becoming 1300 hours and six p.m. becoming 1800 hours. The military also changes the words associated with each time. For example, two a.m. becomes 0200, and is spoken as "Oh two hundred hours," while two p.m. becomes 1400, spoken as "fourteen hundred hours." Quite simple and straightforward. If you use it all the time, or even part of the time, you get used to it.

Peter introduced his own version of military time— never using it in writing, just spoken. He asked Stoner for some time off because he had an appointment at "0 fourteen hundred o'clock in the afternoon." Stoner never corrected him. Later, Peter pressed it further with "half past 0 fourteen hundred o'clock in the afternoon." No reaction from Stoner. Same thing with "at a quarter to 0 eleven hundred o'clock in the morning," no reaction. He tried it with the Major and got only a smile. The Colonel had to bite his lip, but he, too, didn't react to Peter. They found it innocent, humorous, and creative, but not worthy of an early exit to Minnesota.

Peter tried something else, another game in the office. After distributing sheets of paper to each of the enlisted men in the office, excluding Stoner, he asked us to write the names of all the states on the paper. He gave us fifteen minutes to complete the test. Standing at his desk, in front of the Major and the Colonel, he watched us. The Major called Ray over and spoke with him. Then the Major leaned over to the Colonel and said something to him. They both watched Peter.

The exercise proved to be a challenge. There are two ways to do it. One is alphabetically, the other is to start with Maine and go up and down across the country. When I did it, both ways resulted in only forty-eight states. I looked at Andy, who appeared to be struggling. Looking up, he saw me and said, "There's two I can't get." That made me feel great. I considered Andy as one of the best minds I'd ever known, and he was stuck on forty-eight with me. I asked him which two he couldn't get, and as the words exited my mouth, I realized how stupid the question was. If he knew which two he didn't get, he'd get them. To my surprise, he answered me. "Oh, I got all the states, and I got the capitals too. I did the senators, and there are two I can't get."

"You wrote all that down," I looked at my watch, "in twelve minutes?"

"Did some abbreviating, but I know what they are."

"And you can't get two of the senators?" I said, no longer his intellectual equal.

"Yeh, they're new, both just got elected in November. One is the Republican who beat Carlson in Kansas. The other is a Democrat who defeated Long in Missouri."

"You keep that kind of crap in your head? Sorry, man, but that's lame."

"Dole. Dole's the guy who beat Carlson. Didn't remember until you said lame. Dole's got a bum arm he hurt in the War. Yeh, I keep all that in my head. I figure when I finish law school, I might try being a lobbyist for one of the big companies. If I'm going to be a lobbyist, I have to know all of them. If I decide to work on the House side, I'll have to know 435 Congressmen. Well, not all of them. The way they do things with lobbyists is to focus on one party, so I'll only have to deal with half of them."

"That's still a lot, and they change every two years."

"Only part; one third of the Senate is up for election every two years, it's the House that gets elected every two years. It's not much when you consider all the people who follow major league sports. They know all the players. Look at the folks who follow college football. Most of those players are only around for two or three years, then, it's a whole new batch of players."

"Still, it seems like a lot."

"Look at Teddy, who's an accountant. Teddy's an idiot, but I'll bet he can do all the numbers, all the way to a million. I'm only dealing with a hundred senators, and I can't remember one."

I started to argue, but saw he had been pulling my leg with the million-number bit. The sports reference was on target though. "You're right. I followed the football Giants and the Yankees as a kid—"

"Eagleton. That's the guy who beat Long."

"Was it the football thing I said? The Giants reminded you of the Eagles, and Eagles reminded you of Eagleton?"

"No. I just thought of it."

Peter came around, collected our tests, and declared Andy the winner with a loud flourish. Ray also listed fifty states, but had listed Alabama and Kansas twice, so he only had forty-eight. Peter declared me second., saying I didn't try to cheat by saying I had fifty. Second to Andy in anything involving the brain was good. I saw the Major and the Colonel later compiling lists. Looked like they wanted to play Peter's game, but off the radar, out of sight of the troops.

During the game and the announcement of the winners, Peter made a point of being near the Major's desk, guaranteeing awareness of another game in the office. The Major didn't respond as expected. When everyone returned to their work, Stoner pulled Peter aside and congratulated him for keeping the men sharp with a current events quiz. Stoner ended his praise with, "Always good to keep up with what's going on in the world."

With Andy's ETS getting closer, I dealt more with Nettie. It became more of a courtesy than a real collaboration. Officially, we were there to help her, but in reality, we did most of the work while she coasted. So, my meetings with her became gossip-swapping sessions.

As I told her a story from Basic Training one day, she giggled. Nettie had a peculiar giggle. Oh, it was the normal "tee-hee," and had the right sound and tone for an overweight, post-menopausal, former active-duty Marine; but she made it peculiar by putting her hand to her face to cover the giggle. If she had been thirty years younger and a hundred pounds lighter, it would have been cute, but she wasn't, and it wasn't. Knowing I had said nothing funny, I asked her why she was giggling. Taking a little gasp of air, she relaxed and told me the word "dude" was funny. I used the term to describe a couple of the characters whose names I couldn't remember.

"Every time I hear a guy referred to as a 'dude' I laugh. When I was in the Marine Corps, if we called a guy dude, it meant he'd been circumcised with pinking shears."

"Why would you say that?"

"Dude meant we thought he might be a fairy." As she told me this, her giggle degenerated into a full belly-laugh. She was just the right age and just the right weight and size to do the belly laugh justice.

I, however, reacted as I think all males would. At the mere mention of anything sharp approaching the genital area, my nut sack shrank, trying to pull my balls up inside me to a less vulnerable position. I don't know what the physiological connection is, but a tingle crawled up my back to make the hair on my neck stand up. Probably an old piece of DNA from the caveman days, when wild dogs would attack our balls and our neck. Nettie saw the shiver. Maybe she saw the hair stand up. She laughed harder, so hard she blew snot out her nose, and had to leave her desk. She ran out of the office. Later, she told Peter she laughed so hard she peed herself, not a lot, but enough. Peter, always looking for an opportunity, now, would say "dude" at every chance to see if he could get her to laugh hard again. He overdid it, and she soon became immune to his taunts.

Peter didn't do his job at all, now. He walked around with a clipboard, and kept his watch at the top near the clip, so he could watch the second hand. The clipboard paper had columns of numbers and notes recorded in the columns. He never put the clipboard down and locked it in his desk at night. It was another game—of that, there was no doubt. Of course, we were interested in finding out what he was up to.

Peter had never been the neatest of soldiers. His shoes were always dirty and in need of a shine. Was he worse than the rest of us? Yes. His hair tended to run on the long side, primarily to accommodate the cow-lick he cultivated. Like most of us, he avoided the barber until told to do so by Stoner.

The place where Peter distinguished his appearance from the others was the condition of his uniform. Peter, like most of the soldiers who worked in the Pentagon, wore the tan khaki summer uniform they issued us in Basic Training. Unlike the synthetic fabric uniforms Les and I wore, it only took one wearing before they looked terrible. On hot summer days, the uniform was uncomfortable, and wrinkled easily. When cooler weather required a coat or jacket, the shirt wrinkled badly, and soldiers showed up for work looking like they slept in their uniform. A few soldiers purchased extra uniforms, or sent them out to a professional laundry, where they got a heavy starch treatment. Peter did neither. He still only had the two originally issued uniforms, and he laundered them himself at home. Washing these uniforms was the easy part; ironing was another story.

When done professionally, the uniforms got soaked in starch and were then sent to an industrial pressing machine. Those who did their laundry at home used spray starch. Peter disliked the scratchy feel of the uniform after starch. Without starch, any creases, particularly those in the pants, disappeared after the wearer sat down once. Peter avoided laundering his uniform because he felt he didn't get it dirty; it didn't smell if he wore it for four or five days. But mostly, if he washed it, he had to iron it, something he didn't like and wasn't very good at.

With Peter in the home stretch in his race to Minnesota, he neglected his appearance even more. His uniforms became even messier. Finding a crease that wasn't a wrinkle was a chore. It looked like he put his uniform in the laundry basket with wet towels, let it percolate overnight, and then shook it out in the morning before putting it on. One morning, he looked like he had even forgotten to shake it out! When he walked in, the Colonel looked at the Major and gave a head nod. The Major, in turn, gave a head nod to Stoner, who got up and followed Peter to his desk where words were exchanged., Stoner stern, Peter playful. Stoner made an emphatic finger point to Peter and returned to his desk.

Seeing it all, Andy decided the time had come to talk to Peter and invited him to coffee at Ground Zero. Andy asked me to join them.

"Peter, we all know you want out of Washington, and when I say all, I include Stoner, the Major, and the Colonel." Andy had his serious face on, but spoke in a soft and kind manner. "Word has it you're going to be out of here before the end of the summer. I'm short, so what you do doesn't affect me. If you continue to screw around with the military side of your responsibility, you can kiss your transfer to Minnesota goodbye, and tear up that requisition card under your blotter. That's your loss. What I'm concerned about is what your behavior might do to the rest of the guys in the office. If you keep screwing up, like with the sloppy uniform thing, the brass will come down on everyone. Word is getting out on you. That may not mean much to you because all you want is to leave. But let me point out something to you. With everyone watching you, especially this uniform sloppiness, word will get up to the Major at HQ Company. Remember, you're still under his control."

Peter squirmed and looked concerned. He had forgotten that little twist in our assignment here: "assigned to HQ Co, USA, Fort Myer, with duty station the Pentagon."

Andy continued talking to a more attentive Peter. "He can drag you up there for disciplinary action. If you check the requisition card you have squirreled away, you'll see it requires a clean disciplinary record. You may think a sloppy uniform is a stunt to get you out of here, but that same sloppy uniform can keep you from getting to Minnesota. You may wind up leaving this office, but wind up at HQ Company as a general duty soldier, and my friend, there is nothing any of us here can do about that." Andy reached out and touched Peter on the shoulder before continuing. "Peter, we want to see you and your family back in Minnesota. So, straighten out for them."

Peter seemed moved by the thoughts behind Andy's words. He promised to straighten out, and said he would work on his uniform that night.

Peter worked on his uniform, as promised. The next morning, he showed up in a fresh uniform, with all the brass shined and correctly placed. His shoes were shiny—not patent-leather shiny, but black with a gloss. His shirt had been ironed. The pants appeared freshly laundered and ironed, with sharp creases on the legs, both back and front. The only problem was, the creases were inside out!

"What the fuck?" Ray said, loud enough for the entire office to stop and look his way. It was the first time in months I'd heard anyone drop the "f-bomb," and it surprised me. While it had been a regular part of our vocabulary early in our Army career, its use all but disappeared. It hadn't been necessary, until now. Peter stood in front of us, the us being all the soldiers, the Major, the Colonel, and Stoner. Nettie wiggled her way to the front of the group gawking at Peter, waiting to hear what he had to say.

"I really wanted to show you guys"—he looked from the Major, to the Colonel, to Stoner—"I could be a good soldier. I've never been a good ironer, and my wife is even worse. She always buys wash-and-wear clothes. So, I called my mom and asked her how to iron. She used to iron my dad's work clothes. He was a mechanic and wore a freshly ironed uniform every day. She told me she always found it easier to iron from the inside out. That way, if the iron got too hot, any scorching would be on the inside. She did most of the hard ironing on the inside, so when you flipped the clothes to right side out, you just put the creases in and smooth everything else out. I did it that way, starting with the shirt. When I finished on the inside, I turned it and did the finish on the outside. I used some spray starch, and it really looked good. It still does," smoothing down the front of his shirt and turning to show everyone.

"Then I started on the pants. I turned them inside out and ironed them like I did the shirt. Boy, they looked good. My girls got into a fight about something. My wife was working, so all the parenting fell on me. By the time I got them settled down, it was time for their baths, then a story and tuck them in. When I came out of their room, I was exhausted. I saw the pants and realized I had to finish them. So, I grabbed the starch and put the creases in them, nice and sharp. I used a lot of spray starch. Folded them and went to bed."

"My wife got home late last night, so I got the girls their breakfast so she could sleep. I was cutting it close already when I started to get dressed. I put on the shirt and it looked nice. Then I grabbed the pants. 'Oh, my God' was all I could say. I had ironed them inside out. I didn't turn them back the right way after I put the girls down for the night. But now I didn't have time to do anything about it. If I wanted to re-iron them, I'd have to wash them again, and I didn't have time. I pulled the other pair out of the hamper, thinking I could press them a little and get by for the day, but they had a stain on them from something the girls had on their clothes. I was stuck, so here I am."

Nettie asked, "Didn't you see the pockets were on the outside when you ironed last night?"

With an angry, sympathetic, pathetic look that said, "You fat, stupid, boozer, of course I didn't see the pockets on the outside," instead he simply shook his head No. He looked at the rest of us for sympathy.

No one spoke. I laughed. Everyone's attention turned to me.

"Yeh, laugh, Jonathan. It would be funny if it happened to someone else, but I hope I wouldn't laugh."

"I'm sorry, Peter. It is funny. The same thing happened to me in AIT. I was on guard duty. The soldier who scored the most on the inspection beforehand got a three-day pass. I spent two days spit-shining my extra pair of boots. You could shave in them they had such a mirror finish. I set one set of fatigues aside and ironed them three times. I shaved the side of my head up to my hat line, a white wall. When it came time to get dressed, I did it carefully, not wanting to put any wrinkles in my fatigues. When I was all dressed, I went to put my belt on, and there were no belt loops. I had ironed the pants inside out, just like you."

Those not laughing at me, were smiling. The heat was off Peter. We returned to work, leaving Peter standing with his trousers creases inside out.

As I left to return to my desk, Nettie smiled, "Nice going, dude."

Peter stayed in the office all day, leaving only once for a trip to the men's room. Stoner brought him lunch at his desk.

Remembering that guard-duty embarrassment also brought back other memories of that night. After being called a dozen synonyms for stupid, and enduring the disappointment of not getting the three-day pass, I remembered the feeling of loneliness and despair as I walked my post. At 2200, they play "Taps." You can hear it all over the post. It brought on desperate feelings as I walked my guard post. I thought they'd send me to Vietnam to die. I felt a past that was lost, and in the future, I would soon be lost, and then forgotten.

But now, those were just bad memories. I would be out of the Army and back in the world in a little over a year.

CHAPTER 31

"Recruiter Loses, I Win"

Not long into the year, I learned Mike Allen, my 902nd liaison, and I had had been assigned to a joint task force with representatives of the Air Force and the Federal Aviation Administration. Neither of us were active members of the meetings but had been assigned to monitor their deliberations about airline safety.

The joint task force meetings proved an interesting diversion, not taking too much of my time. As observers, Mike and I had no assignments, took no notes, and didn't have to brief anyone afterward. The agenda focused on ways to prevent and respond to airplane hijackings.

While not a problem in the U.S., these terrorist attacks were increasing at an alarming rate in Europe. El Al, the Israeli airline, implemented total baggage searches, as well as body searches. The Air Force and Federal Aviation Administration were forced to the table because each thought they had jurisdiction and were duplicating efforts. The Air Force claimed they should be allowed to shoot down a hijacked airplane because it was a potential weapon. FAA officials offered a more cautious preventive approach, more like El Al. The airlines were lobbying both sides, the Air Force for the obvious reason: they didn't want their airplanes shot from the sky. At the same time, they were urging the FAA to study the problem longer, because they felt searching everything and everybody before a flight would cause an uproar from the inconvenienced passengers.

In April, Nixon announced the withdrawal of another 150,000 US troops from Vietnam, and days later, he announced U.S. and Vietnamese troops had entered Cambodia, attempting to disrupt Viet Cong staging areas.

Andy was due to get out of the Army in two months, the week of July 4, his day of independence. Like Peter, he was already out the door, mentally.

"Have you checked on that recruiter recently?" he asked one morning.

When I checked the books, he had been selected for Vietnam.

"He was so good to you, with you winding up here with the job of a lifetime. You should give him a heads-up," he said with a tone of sarcasm perfectly suited to him.

After I checked the file cabinet, I took his card and made the call. "Sergeant. Specialist West here in OPO. How are you doin', good buddy?" I said with unguarded sarcasm.

"I'm doin' good, buddy. What's up?" he responded with unguarded skepticism.

"When I checked the files this morning, I ran across your name. You haven't got the orders yet, but you're going to Vietnam in about two months. Thought you might find a head's-up helpful."

"Hey, buddy. Yeah, thanks, but this isn't a good time for me. The wife just got promoted in her job, and the money is great. Both my kids are moving up in school in the fall; my daughter is starting high school. I can't believe it, and my son is starting middle school. They're both excited. Been thinking it would be good to stay here another year. What can you do for me, buddy?"

"Sounds like you have an exciting year coming up. Let me check the card again. Yup, I was right. Sarge, you're in luck. The IBM card is just for you. The family doesn't have to go, so they can stay there. Your wife can enjoy her new job, and the kids can go to the new school. There's nothing on the requisition about them having to go with you. I'm glad I could help you out on that."

"West. My family can't stay here without me. They can't afford it."

"I don't understand, Sarge. Your wife got a promotion, and you're still in the Army. It's not like you take a cut in pay to go to the 'Nam. You get more money for being in a combat zone, right?"

"Yeah, but—"

"And it's not like you're going to get an apartment off post in Vietnam. Golly gee"—Andy laughed when I tried to sound like Gomer Pyle—"you can have most, if not all, of your pay sent home. They might wind up with so much extra money while you're in 'Nam your wife might ask you to extend." I enjoyed this, because now he had the questions, and I had the ridiculous answers, perfect role-reversal.

"West, you know what I mean. Ain't about the money. This is an important time in my life. I can't go now."

"Sarge. Let me tell you a little story. Once upon a time, a young man got drafted. He was scared and didn't want to die in Vietnam, so he went to a professional soldier, a recruiter, for help. The recruiter lied to him at a very important time in his life. When the young man complained about being lied to, do you know what the recruiter told him?"

Jenkins didn't say anything. Andy was smiling at me. I asked again, "Do you know what the recruiter told this scared young man?"

Still, no answer.

"Let me make it easier, Sarge. Do you remember what you told me? Allow me to remind you what you told me. You told me to stop my fucking whining, because I was in the Army now. Do you remember that, Sarge buddy?"

Again, no answer.

"Guess what I'm going to tell you, Sarge? I'm going to tell you to stop your whining. You're in the Army. You're a lifer."

"So, nothing? No help?" he said, sounding crestfallen.

"Well, I told you without profanity, except for when I quoted you. That should count for something. Besides, I already gave you more help than you gave me. I told you the truth. Orders should reach you in a few days, Sarge. For the record, I had nothing to do with making this assignment. Wish I did. You just showed up on the normal rotation. You've been there for over three years."

The recruiter hung up. After I placed the phone on the cradle, I looked at Andy. "Too much?"

"No," he said. "Revenge is highly underrated! That was just about right."

"But I don't feel any revenge. I wasn't responsible for any of this. All I did was give him some facts a few days early."

"But he thinks you're taking revenge. Isn't that enough?"

"No. I mean, yes, he caused me some grief, but I came out of it okay. Chances are he screwed other guys. He should pay for them, and this wasn't enough for them. The problem is, I don't know how to make him pay for them."

"So, make him worry."

"How?"

"Talk to Suchanek. He might have something up his sleeve."

Phil Suchanek was my counterpart in MACV (Military Assistance Command, Vietnam) Personnel at Ton Son Nhut Airbase, the major transit point for U.S. soldiers travelling to or from Vietnam. For the past eight months, I got a call from Phil on the AUTOVON system. Phil dealt with all the soldiers in Vietnam who had the 7, 8, 9, and zero series MOSs. Unlike me, he had responsibility for all grades, including the Sergeants up to Command Sergeant Major. Unlike me, Phil handled the soldiers when they entered Vietnam, and again when they left country. Phil had a larger work load than I did, but he worked twelve hours a day, seven days a week. I had worldwide, not just Vietnam, so we were probably even. Also, I didn't have to deal with the occasional mortar attacks from Charlie, which gave me a huge leg up.

Phil, a Polish guy from Chicago, had a large family who worried about him, and a girlfriend he worried about. At least once a week, after we finished our business for the day, I'd transfer his call to either his mother or his girlfriend. Here's a guy half a world away, and he's talking to his family every week. He was in Vietnam, and talking to his mother more often than I was. I can't imagine how happy that made them. Most families would look forward to a letter every day. Some guys just sent empty envelopes home, so the family knew they were still alive. The problem was, the letter only meant their loved one was alive a week ago. But Phil's family talked to him in real time. Can't imagine a war where that would be possible.

I had no problem talking to Phil about the recruiter. He already knew a lot of the story from our previous discussions. It didn't seem unusual, but on reflection, I was talking to a stranger like I had known him all my life. Funny, how war can throw people together. It gave me a sense of what the guys in the trenches shared. They had much more on the line, and depended on the man alongside for survival. This gave me a very small taste of that sense of brotherhood.

I told Phil the Sarge was coming over, but I could only estimate the arrival date. I provided him with all the information from his IBM card, and he said he'd take care of it. He didn't know what he'd do yet, but he'd think about it. I emphasized I wanted no harm to come to him, just worry and frustration. I told him I couldn't stand the bout of conscience I'd get if something bad happened to him. Phil asked me to define bad, and I said, no physical harm, no death. He understood.

Andy was disappointed he'd be out of the Army by the time the Sarge showed up in Vietnam. He said he'd call me to follow up. Like most promises made in the Army, he never did.

The beginning of May brought student protests against the movement of U.S. troops into Cambodia. College campuses around the country, including in Washington, D.C., had to shut down because of the disruption. The most tragic occurred at Kent State, where National Guard troops, brought in to control the situation, shot four students. All the soldiers in the office had been in school before we came into the Army, and could identify with the students. At the same time, we were soldiers, and many of us tried to get into the National Guard to avoid the draft and the possibility of going to Vietnam. Had we been successful, we might have been in a similar situation.

News of Peter's return to shabbiness was interpreted as a sign he was getting ready for more bizarre behavior. Guys in neighboring offices approached him in the hallway with schemes. Some were extensions of college pranks they enjoyed in more carefree days, while others reflected the anti-war sentiment generally found only outside the Five-Sided-Paper Factory. Peter ignored any suggestion of anti-war or anti-administration activity. As a patriot from the Midwest, he turned a deaf ear to anything left of President Nixon. He listened to the pranks and schemes with more than curious interest. But, to his credit, he dismissed each, and returned to his clipboard project.

During this time, maybe because Peter's actions became more visible outside the office, and perhaps a threat to the internal security of the Pentagon, I saw Mike Allen more frequently. He had simple questions I could answer during a short phone call, but he preferred the face-to-face contact.

The end of May, Andy removed his IBM card from under his blotter and returned it to the file cabinet. Now, the Army knew he was in the Army, so they'd be able to process him out. The Finance Center at

Fort Benjamin Harrison could authorize his final pay, and authorize a moving company for his Permanent Change of Station, or PCS, to New York. Two months later, he'd be back in law school, starting the fall term. That he'd return to law school had never been in doubt.

As the time to leave Washington neared, his interest in any part of the political process waned, as did his interest in becoming a lobbyist. His future, he now felt, was corporate law, specifically, corporate law in New York City. The culture and sophistication of New York became more and more a part of his conversation.

At the end of June, on the Thursday before Andy left, I noticed Teddy in a frantic flurry of activity. With his desk cluttered with more paper than usual, his coat jacket open, he banged away on the fifteen-character adding machine he kept on his desk. A Major I hadn't seen before stood hovering over him. It was the time of week where Teddy provided the troop strength data for the secretary of defense's Friday afternoon press conference.

Each Friday, the SecDef provided the number of soldiers still in Vietnam, for comparison against the plan they had revealed for the "Vietnamization" of the war. The number was much anticipated by the press and the public each week, as a measure of Nixon's commitment to end U.S. involvement in the increasingly unpopular war. As I learned later, the number as it appeared from the secretary's briefing also got the attention of others.

Teddy rifled through rolls of computer printouts, banging on his adding machine, opened drawers of his desk, banged on the adding machine, pulled the crank, stared at the number, then started over again.

Andy watched with as much interest as I did. "Why the intensity today?" I asked.

"The new numbers for withdrawal came out at the beginning of the week. The press call it Phase II. If they don't hit the target, there'll be hell to pay."

"And they're relying on Teddy to provide that number?"

Andy looked up from the paper on front of him. "Oh, he'll get the right number. Teddy knows what the target is, knows we shouldn't hit it on the nose, and knows we can't be over the target. The fifteen-column adding machine is his random number generator."

"What?"

"The number this week is supposed to be around 222,000. Teddy knows the exact number before he starts on his adding machine— let's say it's 222,123 this week. As he looks at the computer sheets, he's randomly banging away on his adding machine. When he stops, he looks across the readout for 222. If he finds it, and the next three digits after 222 are less than 123, then 222 plus those three numbers are the number for the week. If no 222, he starts over again. Today, he's frustrated, which probably means he can't find a string of 222 with less than 123 after it."

"Makes sense—there's only a twelve percent chance he'll find 123 or less."

Andy looked at me, a looked that masked thought, then his eyes lit with agreement. "Never looked at it like that. You're right."

Teddy had his eureka moment, almost collapsing at his desk after writing the number on the Major's tally sheet. Teddy pulled the paper tape from the adding machine and tossed it into the waste basket. Exhausted, he left his office for a break.

While he was gone, my curiosity got the better of me, and I retrieved the balled tape, uncrumpled it, and noted the number, especially the last three digits. The number he circled was 118. On Saturday, I checked the newspaper for the number reported by Mr. Laird. The first three numbers were 222, but those weren't what I was interested in, just the last three. The secretary reported the number of troops remaining in Vietnam, and the last three digits were 118. Son of a bitch! Fake news.

Monday morning, Louise, the lady who handled the language school assignments, came to work excited and happy. Announcing she had hit the bookie's number on Friday night, she bought breakfast for her buddy, Thelma.

"Congratulations, Louise."

"Thanks sugar—a nice payday, five hundred big ones."

"That's incredible."

"Yup. Been playing my birthday for two years: January 18, one, one, eight, and it paid off."

"You play every week?"

"Yup, place my bet Thursday on my way home from work."

"If I give you some money next week, can you buy me a number?" I had a hunch. To check it would only cost me a buck.

Louise looked perplexed, "You don't know any bookies?"

"Not here. Back home I knew two, Moo-Moo and Big Frank."

She agreed.

Andy's last week turned out to be a celebration of sorts. Joey Heatherton gave a performance in the courtyard surrounding the Ground Zero Café. The sexy entertainer was married to Dallas Cowboys star receiver Lance Rentzel, who the previous fall had earned the nickname "no pants Lance" after being arrested for exposing himself to a minor. We tried in vain to get her to bring Andy on the stage. Would have been a great send-off for him.

On Thursday, the day before Andy's "wake-up," I retrieved Teddy's crumpled tape from the garbage and gave the number and one dollar to Louise. Soon after, Andy cleaned out his desk and moved everything to my desk. He shook my hand and said, "Time to bust out of this puke hole." As he walked to the door, he shook hands with the Major, the Colonel, and Stoner, then out the door.

He never called again. Nothing really changed. It was like Andy was never there. People seldom spoke of him again. Andy was the first friend in the office to leave since I arrived. Returning to the civilian world should have been a bigger deal, but it wasn't for the guys leaving 1D726.

Monday morning, I started my first day as the guy in charge. When Louise returned to the office after getting breakfast, she handed me an envelope with five hundred dollars in it. "I been playing the numbers for years and hit just that one time. You play once and you hit."

"Just lucky, I guess," I said as I took the envelope. I opened it and took out the cash.

"It's all there, or don't you trust me?"

"Of course, I trust you," I said as I took out two twenties and passed them to her. "Your commission. Thanks."

"You going to do it again?" she asked.

"Sure. Do you use the same bookie all the time, or do you have others you can use?"

"I have a few, why?"

"'Cuz I plan on winning for a while, and want you to spread it around."

And that's the way it went for the next four weeks, until the bookies in Washington used a different source to get the number. I made twenty-five hundred dollars, less Louise's commission, and she bought a new car with the money she won betting my number.

CHAPTER 32

"Peter, Paul, All So Merry"

Peter lasted a week being a sane soldier. Then the need to get out of Washington and back to Minnesota returned, evidenced by the reappearance of the crumpled uniform and the clipboard, beginning the week Andy left. Maybe that's what triggered it. Nobody bothered him much about the uniform. With the summer heat and humidity, everyone who had one of those old uniforms looked shabby by mid-morning. Peter just got a head start on everyone each day.

Peter became more open about his project, claiming it to be of great importance to the Department of Defense, not just the Army. He claimed it was so important he would get an ARCOM, the Army Commendation Medal, in addition to a ticket to Minnesota.

Peter spent a lot of time out of the office. Once out, no one knew where he went, or what he did. The Major warned me again, if Peter screwed up, all bets were off as far as Minnesota was concerned. With Andy gone, I became Peter's guardian angel, a reluctant guardian angel.

I didn't have the time for the angel bit, now that I had a double work load. The only light at the end of the tunnel was, with Andy gone, I could look for my replacement. That was the process. Andy brought me in and trained me as his replacement. When he left, I had to do the same.

With an allegiance to Fort Dix, I called Personnel up there and ordered four clerks from their school. Simple criteria for selection: The soldiers had to fit the composite of all the guys I had worked with. They had to be college graduates, preferably with work towards an advanced degree, and they had to type. Anyone drafted out of law school should be on the list. After a few days, they called back and said they had four candidates. I told them to send them down. I'd have the first pick, and if they didn't fit my needs, I'd send them to other openings around Washington. Then I'd ask for more candidates.

While I waited for the new batch of soldiers to arrive, I got news Peter was preparing his report, based on research done while he was carrying his clipboard around. Peter worked at his typewriter for two days, typing a single copy of the report. He refused to use carbon paper, because he made too many typing mistakes and correcting the carbons took too much time. Rather, he corrected with Wite-Out, using th bettr part of a bottle for his document. If he needed additional copies later, he'd use the copy machine down the hallway. At noon of the third day, he presented his report to Stoner. He asked Stoner if he wanted a briefing before he read it.

Stoner asked, "Is it in English?"

Peter responded, "That's affirmative, Sergeant Major."

"Good, I should be able to read it just fine."

When Stoner finished, he gave it to the Major. The Major gave it to me, saying, "It might be time. This is crazy, but if we don't let him go soon, he's going to do something really crazy," waving the paper in front of him. "He wants out so bad, I'm afraid he'll show up wearing a dress. If that happens, he'll get a Section 8 discharge. The little guy'll be home in Minnesota for sure, but he'll be labeled a crazy for the rest of his life."

I had too much to finish before the afternoon call from Vietnam, so I set it aside to read the next morning.

Peter's treatise was both funny and disturbing. The information he collected showed the personnel at the Pentagon, both civilian and military, wasted taxpayer money, wasted time, and wasted water. This waste was caused by spending too much time going to the bathroom during working hours.

Peter started with a justification of himself as the typical American, and therefore the model for normal. He described himself and his background as from the upper Midwest, of average height and weight, having normal eating habits and normal activity. He used himself to set the baseline as one bowel movement a day, in the morning before he left the house. The baseline for liquid consumption during the day was coffee in the morning, one cup, a soda break in the afternoon, along with a beverage at lunch. This required him to take two pee breaks during the day, one of which occurred during the lunch break. He timed each of his bathroom breaks and estimated the amount of water used to flush the toilet and wash his hands. These were all used to establish the base line.

The scary and disturbing part came when he recounted how he collected information for the general population working in the Pentagon. He started by following people as they left our office, recording the time away from their desk. Taking it further, he followed the men into the men's room and recorded their activity, putting a check mark beside each incident under #1 or #2. He recorded the number of flushes and time spent washing hands, for an estimate of water usage. He acknowledged there may be a sampling bias in his data, because he only used people in our office. To offset this bias, he loitered around other bathrooms, recorded the data, and then followed these strangers back to their offices.

Peter took this data and extrapolated it to the population of 35,000 people who worked in the Pentagon. He concluded that bathroom breaks above the baseline defined as normal, accounted for a loss of over 16,000 man-hours per day. He also calculated wasted water and paper products. For the paper products, he postulated women used more than men, because men could shake it dry. At the end, he recommended the Secretary of Defense issue a mandate limiting bathroom breaks to that established for the average American, Peter.

I put the paper down on my desk and sighed. The Major had given me the paper to read and made it my problem to solve. Time to step up to the plate. I grabbed the paper and went to see the Major. As I approached his desk, he motioned me to the chair inside his cubicle.

"So?"

"Well, Peter's not a scientist. The baseline comparison is his own experience, so it makes the whole thing anecdotal, not science. Data collection—"

The Major interrupted, "I know his crapper study is a pile of shit. What I want to know is whether this is part of his scam, or is he going nuts on us?"

"It's part of his scam. He started this thing a while back. We saw him with the clipboard, but we had no idea what he was doing. Things aren't going well at home, and he's feeling the pressure. We shouldn't wait until August; we should let him go now."

"Stick around a few minutes after work tonight. I know you ride with the Sergeant Major and Ray, but they can wait. Stoner knows, and Ray will find out next. Grab the card for Minnesota from under his blotter and cut him orders, effective tomorrow. I'll sign them first thing in the morning. We'll cut him loose immediately. Do you see any problems?"

"First, I'll need to grab his IBM card and put it back in the system. Without the card, he doesn't exist, as far as the Army is concerned. When the orders leave this office, they'll bounce around looking for him. Then the orders will bounce back here."

"Can you handle that?"

"Yes, sir. Had to do it for Andy last week." As I said this, he gave me a questioning look. Before he asked, I said, "Sir, no one in this office exists. Our IBM cards have disappeared from the system. We control them." He smiled, shook his head in a disbelieving way, then gave me a head bob dismissing me.

Both cards were under Peter's blotter. Once I prepared the orders, I put his personal IBM card in the rack across the hall, where the Frito Bandito used to reign. The card and Peter would be back in the system before midnight. Peter was back in the Army database, and we could send him to Minnesota.

The Major signed the orders in the morning, called Peter over, and handed them to him with no emotion on either's part. Peter cleared the personal things from his desk and came over. He shook my hand and said, "Time to bust out of this puke hole. See you around the campus," and he was gone.

No one ever heard from, or of, Peter again. No one bothered to check if he made it to Minnesota. He didn't send us a post card. It was like that most of the time. We were close when we worked together, and considered ourselves a group of lucky guys to have wound up safe in the Pentagon. But once it was over, it was over. No farewell parties, no gifts, no nothing—just gone.

The four candidates for my replacement arrived from Fort Dix the next day. Three of them didn't feel right. While qualified, there was something about them that made me feel I couldn't work with them. It was my first experience "hiring" someone, and I honestly didn't know what to do. The only thing I could relate it to was meeting a girl for the first time. With a girl, it's the superficial that first gets you; nice smile, cute laugh, a great body, unique voice. Then you ask yourself the questions. Why does she always have that stupid grin on her face? That laugh is driving me crazy. Does she know what she sounds like? She'll drive me crazy with that body. She's a tease. I can't stand her voice. She sounds like one of the Chipmunks. So it was with the candidates from Fort Dix. All were qualified, so it fell to the basic question of, "Do I want to spend eight hours a day, five days a week, looking across this desk, listening to this man?" For three, the answer was no.

The soldier I selected was a second-year law student from New York. Yes, I realized I was picking my replacement by matching basic criteria to Andy's background. But beyond that, they couldn't have been more different. Paul Wittenstein was tall to Andy's short, and skinny to Andy's rotund, with curly black hair where Andy had no hair. His facial expressions were neutral compared to Andy's perpetual smile. Did I mention he was Jewish? He was. Not that it mattered then, but it would become an issue later.

I started Paul with Nettie, just as Andy had started me. Good for him, and good for her, because with Peter gone, she was now down a man. Did I mention that Nettie didn't like Jews? I didn't know it until she generated a list of things he did wrong. A long list. Nettie presented me with the list and told me he wouldn't work out. The list contained several things Peter had done repeatedly and equally wrong, but that had gone unnoticed or was tolerated. When I pointed out his faults seemed similar to those of Peter, she leaned in close, covered her mouth with her hand, and whispered conspiratorially, "He's Jewish." I expressed surprise, and she nodded her head yes.

"Are you prejudiced against Jews?"

"No. Prejudice means you don't like colored people. The Marines had lots of colored people. I got along with them fine, even had some who were friends of me and Alfie. But we didn't have any Jews in the Marines, so I don't know anything about them except for the stuff you read."

"You mean like the stuff Norman Lincoln Rockwell writes?"

"That picture guy from the Saturday Evening Post?"

"No, the guy from the American Nazi Party."

"No, I never read anything like that."

I teased. "Did you know Audie Murphy was Jewish?"

"No, really?"

While Nettie thought about that, I called over to Paul at his desk, asking him. "Are you Jewish?"

He said, "Yes," with more of a question in his tone than a declaration.

I said, "Nettie thought so. Have you found any good Jewish delis, like they have in New York?" I left Nettie's desk and started back to mine, to continue my conversation with Paul.

As I made my way to my desk, I heard Nettie say, "But Audie Murphy wasn't a Marine, he was Army."

"Not yet, but I'm sure there have to be a few," Paul said, responding to my question.

"Let me know when you do. I love deli. My favorite is hot pastrami on rye. Big kosher dill, but hold the Cel-Ray, you can keep that."

His eyes lit up. "Yeah, Cel-Ray is an acquired taste. What about an egg cream?"

I wrinkled my nose in disapproval. "You're not going to like this, but my favorite beverage with a hot pastrami is a coffee milk shake. I know I can't get it at a deli, but I can do take out."

Paul and I had bonded. My experience with deli food was limited to a few trips to a kosher deli with classmates from New York, but it was enough. Nettie backed off, and Paul settled into learning the job with her as an unwilling tutor.

The only difference in my job now was when I looked across at Paul, working with his head down, the glare was gone. I saw black curly hair, where before I had seen the fluorescent light reflected off Andy's bald head.

Mike Allen returned from a two week leave, having missed one of the joint task force's meetings. We had lunch to catch up. After I gave an update on the nothingness that happened at the last meeting, he mumbled through his mouthful of steamship side of beef, "Peter was a crazy little fucker. Probably best he moved on. He had some crazy ass little schemes brewin' in his head. No telling what he would have gotten into, and how it would have spilled over on to you."

"You know about Peter's schemes?"

"Everybody knows about Peter." Then correcting himself, "a lot of people know about Peter."

"How come a lot of people know about him?"

Mike put his fork and knife down across the top of his plate, wiped gravy from his mouth, then said, "Because I tell them. And it's not a lot of people, just people who are interested in you, the people around you, and how you handle those people."

Perplexed, I asked, "Why are people interested in me?"

"I can't tell you now. Perhaps in a few months. For now, let's get back to Peter. Do you know how close he came to being dishonorably discharged?"

"No." This was a surprise. Peter had done some bizarre stuff, but nothing that warranted a dishonorable. "I think he went over the edge with worry about his family, but I don't know if that would qualify as crazy."

Mike had started eating again, and continued where he had left off about Peter. "I heard about that toilet thing. That was spooky in itself. Hanging around the bathrooms and all. I'm surprised someone didn't turn him in as some kind of pervert. While I was on leave, Peter asked a buddy of mine, who shall remain nameless, to meet him for lunch. Instead of the Ground Zero Café, Peter wanted to go to the Concourse."

"That's unusual?"

"Yeah. Peter said he wanted him to experience the setting and the crowds as he described a plan he had. With a squirrelly look in his eyes, he said he wanted to be remembered for a really big 'caper,' as he called it. He told him he wanted to come to the Concourse at lunch hour wearing a raincoat. Under the raincoat, he'd be wearing trousers that only had legs in them. He'd cut the legs off a pair of uniform pants above the knee, then tie the lower pants legs to his legs. If you looked at him, it would appear he had pants on under the raincoat, but he would be bare ass naked. He would be carrying coins, quarters, about twenty dollars' worth. After he'd picked his time and space to maximize the number of people around him, he'd drop the quarters like it was an accident. He'd squat down to pick them up and hope other people would help him. Here's the sick part. While he squatted there, he planned to take a dump. Then, while everyone busied themselves picking up the quarters, he'd fade into the crowd and disappear, leaving the pile of shit there on the floor of the Concourse."

"That's crazy! Why did he tell your friend?"

"Because he thought he was someone who enjoyed a good prank. He also wanted money to do it. Said he'd do it for $500, and my friend should go around asking guys he trusted to chip in."

"What did he tell him?"

"He said no and walked away and told him he was crazy. If they caught him, they'd put him in jail, and probably give him a Section 8 Discharge, or whatever they give to crazies. Peter just laughed. If he had gotten the money, he would have pulled the stunt and be in jail now"

"Jesus, he was crazy."

Paul and I settled into the work routine without Andy. As expected, I became the "go to" guy for the office. The Major, the Colonel, and Stoner relied on me for the one-off assignments.

There was a lot of pressure from the popular press and the television news to accelerate the withdrawal of troops from Vietnam. Nixon had initiated a steady downsizing of Americans in Vietnam, while turning the responsibility for prosecuting the war over to the Vietnamese. Melvin Laird still had his press conferences on Friday afternoon, during which he announced troop strengths. The nightly television news would report those numbers, but took more interest in reporting the number of dead and injured G.I.s for the week.

Paul started working more independently. I gave him the daily calls from USARV, while I kept the contact with Suchanek in MACV. It was a strange feeling, watching Paul. Nine months ago, I came into this office unsure of my future. With a few lucky breaks, I found myself on permanent assignment here, learning the job from Andy, a job I had no concept even existed a year earlier. Yet, in a short time, I mastered it, rose to become one of the more critical people in the office, and was now training my replacement. Maybe I had overestimated the grandeur of the people and mission of the Pentagon. Maybe I, as mediocre as I felt, stood out because the people and the product of this building just weren't that good. Could I be the one-eyed man in the world of the blind?

But then, I thought of the people around me, dedicated professionals with a sincere reverence for the soldiers as people, not just IBM cards—and I replaced Andy and he certainly had more than one eye. If I could do his job as well as he did it, did that mean I was as good as I thought he was? Was I better than I thought I was? Was the Army a place that matured me and gave me some confidence?

CHAPTER 33

"Shoes and Spooks"

Paul had a head for the regulations, so he learned the job quickly. We had to be familiar with the regs, not because we always worked within them, but to identify those times when someone on the outside attempted to work around them and use us as unknowing dupes. As hard as it might seem, there were people making money on the war. Not the big corporations selling guns and airplanes to the Army, but soldiers, some of them draftees, taking advantage of the circumstances (or worse) of other soldiers.

Throughout history, cooks and supply personnel profited from their positions. Large caches of food and beverage had always been targets for diversion, if the money was right and the risk low. The same could be said for equipment. These were generally considered crimes against the system, the Army. Steal something from the Army and sell it to someone else. "No harm done," the mantra of the criminals. But, often, it meant soldiers did without food, supplies, or medicine. Some soldiers profited from selling favors, and I learned it firsthand.

One day in late summer of 1970, I received a call from one of the top NCOs at Fort Myer. The Sergeant had a nephew returning from Vietnam who wanted to go to Fort Collins, Colorado. Being an avid skier, he wanted to take advantage of the long winters and the mountains to pursue his passion. The NCO told me he would make it worth my while if I made the assignment happen.

When I checked the records, I found he had requested Fort Collins and had gotten it. In fact, if there had been twenty soldiers with his MOS requesting Fort Collins, Colorado, in winter, they all would have gotten it. Not many soldiers wanted to leave Vietnam in December and arrive in Colorado in January. When I told the NCO his nephew had orders to Fort Collins, he thanked me and said he would take care of me. Despite my insistence I had only read a report, I couldn't convince him I did nothing to make it happen.

Three days later, I arrived in the office in the morning to find a shoe box on my chair. In the box, I discovered a pair of "Third Herd" shoes, and a simple, unsigned note saying, "thank you." The shoes probably came from the NCO with the nephew. When I told Stoner the story, he shrugged and told me to report it to the Major. When I told the Major, he said, "Looks like someone thanked you for a favor. Did you?"

"No, sir. Just read it from the report on assignments made last week," showing him the report dated the week before.

"Good, because if you had sold the assignment, we'd have to reconsider keeping you here. As it is, somebody just paid you for a favor they think you did. Enjoy the shoes."

"But sir, it doesn't seem right, if he thinks he bought me with a pair of shoes."

"If that's what he thinks, he'll be back for another favor. When he does, let me know, and we'll handle it then. There's nothing here now."

"What should I do about the shoes, sir?"

"Wear them."

I didn't feel right, but I loved the shoes. I called them "Third Herd" shoes, because only the soldiers in the Third Infantry had them--more specifically, the soldiers who guarded the "Tomb of the Unknown Soldier" in Arlington. The shoes were patent leather—not the patent leather my little sister wore when she took tap dance lessons, but heavy-duty, sturdy shoes that withstood the inclement weather the guards had to endure. They had horseshoe taps on the heels, metal plates in the toe part of the sole, and more metal plates on the insides of the heels. These plates were for effect, not durability. The plates on the inside of the heels provided a sharp sound when the guards clicked their heels together, after making the twenty-one steps from one side of the tomb to the other. The plates on the toes of the sole and the horseshoe taps produced a distinct, sharp, strong sound with each step the guard took arriving or leaving the ceremonial post.

I put the box in my lower desk drawer, opening the drawer to look occasionally. They were nice shoes; more than nice, they were great shoes. I had walked around the Pentagon halls and corridors a lot, and had never seen anyone with such wonderful shoes. Still, I felt bad, because I knew as soon as I put them on, there'd be an NCO who thought he had bought me.

Just before lunch, the phone rang with the NCO whom I suspected had sent the shoes. "West. Did you get my token of appreciation? Had a friend drop them off last night."

"Yes Sergeant, I did, but I have to explain again: I had nothing to do with the assignment your nephew got. All I did was read to you an assignment that had already been made."

"Well, West, I understand why you're saying that. If that's what you want me to believe, then okay. But my nephew, his family, and particularly me, are all very grateful."

"No, you don't understand. I did nothing. You feel I did you a favor, and I didn't. I can prove it. I showed the Major the printout where I got the information. He knows I didn't do anything special."

"You told your Major about the shoes?"

"Yes, because you put me in a tough spot giving me those shoes. The Major wanted all the details, so I told him everything."

"Did you tell him my name?"

"No, but I showed him the printout. Your name and your nephew's last name are the same. It didn't take much to put two and two together."

"What did he say?"

"Said I should enjoy the shoes, and report you if you ever ask for another favor."

"Enjoy the shoes." With that, he hung up, and I never heard from him again.

Now, with a clear conscience, I took the shoes out of the box and put them on. They fit well and looked great. As I walked to the Major's desk, he said, "Nice shoes."

"The NCO in question called, looking for a thank-you. Told him I didn't do anything. I told him I'd report him the next time he asked for a favor. Told him you were on board with all that."

"What did he say?"

"Enjoy the shoes."

"Well, that settles it. Enjoy your new shoes." And with that, I set out to have lunch in my new shoes.

The shoes made no sound on the office's worn carpeting, but once I got into the tiled corridor, they sounded great. Putting taps on shoes was the rage when I was a teenager. The taps had to be securely fastened to make the solid sound. If worn thin, or if a nail came loose, the sound changed from the deep masculine sound to a tinny feminine sound— or at least that's what everyone told everyone else. Growing up near Saratoga Race Track, I remember the thoroughbreds walking from the paddock, crossing the macadam paths, and making that glorious sound. My shoes made the same sound: CLOMP, Ka-clomp!

Heads turned as I walked down the corridor. I took the escalator up to the Concourse Level. On the Concourse, I would share the sound of my new shoes with as many people as possible. I enjoyed watching people, men and women, military and civilians alike, turn around as I approached, looking to see this important-sounding person. Some slowed down or stopped to watch and listen as I passed. The shoes were magnificent. After lunch, I started the return to the office.

The Pentagon, as a concept, developed in the late 1930s. The War Department occupied seventeen different buildings in Washington, and needed to bring them all together. While they farted around, as governments do, trying to decide on the location, the size and the shape of the building, Hitler heated things up in Europe. The government sped up the planning process and began building. The official ground-breaking occurred on September 11, 1941, 911 just like the emergency telephone number. When the Japs bombed Pearl Harbor, Colonel Leslie Groves, the Chief of Construction who would later lead the Manhattan Project, decided construction would run fourteen hours a day, seven days a week. To save on steel needed for ships and tanks, Groves had ramps erected between floors instead of elevators—relatively steep, a grade of eight per cent. By the time I got there, they had installed escalators between floors, but these were set to only go up; going down remained only by the ramps.

Having used the ramps regularly, I never gave them a second thought. This noontime, I should have given more consideration to the ramps. Still enveloped in the "wonderfulness of myself and my new shoes," I noisily made my way to the top of the ramp to descend to the first floor, thinking about how to modify the sound. Should I land heavily on the heel and bring the toe down, so I made two distinct sounds? Should I try to change my walk so both heel and toe hit the floor at the same time, intensifying the sound? What would these shoes sound like on a wooden floor? What would these shoes sound like outside, on concrete? I thought about everything except the steep ramp I approached, a ramp composed of shiny tiles.

Pentagon staff used these wide ramps in both directions. Lines at the escalators got backed up, particularly at the heavy-usage hours, like lunch. The escalators also were out of the way. As with any corridor, traffic stayed to the right in both directions. Handrails along the walls helped those who needed them. I took the path less travelled, right down the middle between those ascending and those descending, giving more people the opportunity to see and hear my new shoes.

The road less travelled proved to be a mistake.

Had I been paying attention, I might have seen the potential for disaster approaching with the ramp, and chosen to descend with some caution. Caution may have reduced my embarrassment. Had I not been so proud of my new shoes, I might have chosen the inside path rather than the center. The inside path would have afforded me the handrail.

But, having chosen my path, not with reckless abandon, but with no abandon at all, I began the descent. After the first step, I was in trouble. The shiny new metal plate on my heel came down on the shiny tile ramp-tilted downward at an angle of at least eight degrees, maybe ten. With virtually no friction between the two, my foot started to slide. At that precise moment, any semblance to walking with dignity disappeared. Reflex took over and the other foot came forward to brace for a nasty fall, only to be planted solidly toe-and-heel at the same time. The toe-and-heel combination provided twice the shiny new metal surface to slide on the shiny tile floor. Now, had I been cautious, this might have been the end of it. I might have caught myself and proceeded more carefully. At the very worst, I might have fallen after taking one or two steps. I might have been able to look at an invisible spot on the floor, and claim it had been wet. Had I been to the far right, I could have caught myself on the railing, and suffered no ill effects at all, except for a momentary quickening of the heartbeat.

None of these happened, as I had chosen the middle of the ramp. When the second foot came down, my autonomic nervous system took over, preparing for a fall. The arms started up in an attempt to gain some balance. With arms flailing, the feet had to fend for themselves, continuing forward and downward, sliding, increasing speed, almost propelled by the flailing arms. They divided the ramp into two sections, with a small level landing halfway down, to reduce the severity of the grade. I saw this landing approaching and relished the idea I could stop this embarrassing descent, move to the railing, and regain a sense of dignity. That proved not to be the case. The narrow landing only slowed my descent. The slowing allowed me to move to the right, but I was still moving. As I slid right, the railing and safety in sight, I was struck by someone on my right. In actuality, I struck someone. The result being the same, it turned me around just as I almost reached the railing, but exactly as I reached the beginning of the second descending part of the ramp.

Now I was sliding down the ramp backward, feet planted and arms still wind-milling. I couldn't see the end of the ramp, but I sensed I was approaching it, and the end of this horrific exhibition. I thought how lucky I had been, grateful to be still on my feet, and, except for the glancing blow that had turned me around, I hadn't collided with anyone. Gratitude should have waited until I stopped, because at that moment, I fell.

People around me offered hands to help me up. Several people asked if I was hurt. Taking the offered hands, I shrugged off the concerns about injuries and said I was fine. When I raised my eyes, there were people laughing. The "wonderfulness of myself" was now only exceeded by my profound sense of humiliation.

When I returned to the office, I took off the shoes and put my old ones on. That night, at home, I removed all the metal plates from the shoes. I would have to be content with the memory of the sound of the thoroughbreds at Saratoga, because all I was left with was the shiny part of the shoes.

As I learned more about Paul, I found him to be a complex individual, an intellectual as opposed to an athlete. Paul admitted he had never tried to play any competitive or team sports, ever. He grew up in New York City with educated, religious parents. Being religious and Jewish meant he couldn't take part in any sport with games late Friday afternoon or evening, and certainly nothing that required playing on Saturday. This eliminated all the major sports and most of the minor ones. When he got to college, he realized he wasn't as religious as his parents wanted him to be. He also realized he was ill-prepared to participate in any team sports. To his surprise, he found watching sports fascinating, and became an avid fan of his college basketball team, NYU. As he neared graduation, his interest in college basketball waned and he tried professional basketball, attending games of the New York Knicks. He soon lost interest, or as he put it, he didn't identify with tall, black men making outrageous sums of money. He turned to professional hockey because the season extended from September through April, the longest of any professional sports. This allowed him to be a fan of one sport for most of the year. He chose to be a Philadelphia Flyers fan because he didn't like the Rangers fans. With Philadelphia close to New York, he picked up the games on the radio, his preferred way of following his sport. Philadelphia was also close enough he could attend the occasional game. When Philadelphia travelled to New York to play, Paul made it a point to attend the games for the pure joy of cheering for the opposition. He delighted in being in Washington because he could make the drive to Philly for a game if he wanted.

Because he had grown up in New York and had lived in dorms at NYU, Paul knew a garden apartment in Arlington wouldn't suit him. He had most recently been a student, and after his military obligation ended, he would return to NYU to finish law school. He wanted to stay in touch with the student life, so he answered an ad he found in the George Washington University school newspaper, advertising for roommates. It turned out to be a good fit for him. He left his car on Fort Myer and would not have to compete for parking in Washington, yet still have it available. Travel to work for the most part was an easy reverse commute by bus, out of the city to the Pentagon.

Paul's background security check continued after he got cleared to work at the Pentagon. He needed a Secret clearance, and that is what he had, but the file was never closed. The FBI kept digging into his background. Adding to this was the suspicion Paul's new roommate sold marijuana to soldiers in the Washington area. Because Paul was a new contact of his, they added Paul to their watch list. When they started watching Paul, they found he regularly left his new shared apartment with a gym bag in the morning. Occasionally after work, he picked up a car from the Fort Myer parking lot and drove to Philadelphia, where he would get lost in South Philly near the sports complexes, a hotbed of drug activity. Early the next morning, he would return his car to Fort Myer and go to work.

I realized this when Mike Allen of the 902nd asked me to create a background to plant one of their spooks. The FBI and the 902nd were trying to get information on a suspected drug distributor at GWU. When Paul showed up in their surveillance, he became the means to infiltrate the group, but he also became a suspect. They wanted to assign the 902nd agent to an office on our corridor, befriend Paul, and hopefully work his way into the drug dealer's inner circle. The bonus for the FBI was the possibility of capturing Paul and his contacts in Philly.

When he explained this, I was shocked to think Paul was involved in drugs. The more I thought about it, the more likely I thought them wrong. Paul had finished his first year at the NYU School of Law. During my interactions with him, he gave no hint he would do anything to jeopardize his return to school, his ultimate graduation and licensing. His goal was a career in international law. Paul came from money, and his parents financed the lifestyle he couldn't afford on a PFC's salary, and his trips to Philly were to see his beloved Flyers. The likelihood he was a marijuana mule for some college kid at GW seemed more than unlikely.

If Paul enjoyed a joint once in a while and his timing was wrong, he might get caught up in something that could ruin the rest of his life. I shared none of my reservations with Mike Allen. My obligation to the 902nd prevented me from telling Paul.

The planted agent did in fact befriend Paul. He expressed an interest in the same things Paul liked: the nightlife, as it was on the GW campus, a career in law, and a love of hockey—although his allegiance was to the Hershey Bears, a minor league franchise in Pennsylvania. The guy became a regular visitor to Paul, going to lunch, stopping by for coffee breaks, and talking all the time about hockey and the girls at GW. I commented to Paul he and the new guy seemed to have a lot in common.

"Yeah, he's all right—a little forceful about getting to meet my friends. Dude keeps asking me if I know someplace where he can score some grass."

"Do you?" I asked, then added, "Don't say dude around Nettie."

"Not here. At NYU, I could get it. Like every college kid, I smoked a joint once in a while, but when I got into law school, I weighed the risk and swore off. I don't need a bust for a joint on my record. It'll kill my future, not to mention what it would do to my parents. Ain't worth it. I've seen guys, and some girls, just throw away their future over a joint. It's harmless, no worse than alcohol, but the law is the law."

I was convinced the 902nd was barking up the wrong tree with Paul, but he was still vulnerable if he introduced his new friend to the suspected dealer at GW. If the dealer got busted, Paul would be sucked up in the vortex of shit that followed. I didn't know what to do, until one day Mullaney made one of his cryptic comments. After seeing Paul leave for lunch with the 902nd spook, he said, "They make a cute couple." Bingo, I had my solution.

"You don't think?" I asked.

"About what?" responded Mullaney.

"What you said about them being a couple."

"Don't know. They seem pretty tight, but I never really thought about it. You know Paul better than I do, what do you think?"

The ball was back in my court, exactly where I wanted it. "I don't have any doubts about Paul. He seems to have a healthy appetite for women, from what he says. The new guy, I don't know. He seems to have an unusual interest in Paul. They say you can't always tell, but he doesn't look like one. He seems to touch Paul a lot, now that you've pointed it out. You know, the shoulder, mostly on the shoulder. When he comes over to talk to Paul, he always stands alongside of him. I guess he could be."

Mullaney picked up and fed off my lead. "Paul should be careful, then. The last thing he needs is to go back to law school being known as having to wear white socks in the Army." Having to wear white socks was the punishment if they found a guy to be queer. It wasn't announced as a punishment, just an identifier, so others knew. Then folks could associate with or avoid the white socks. I never knew if it was true, but the legend was real. "Maybe, you should talk to him."

"Not me, he works for me. It should come from someone independent, like you."

"I don't care one way or another about Paul. Don't really know him; but in this office, he's my brother, and he is a fellow law student. I'll come up with something."

Paul came back from lunch alone. We got talking about plans for the weekend, and Paul mentioned he and his new friend would spend Friday night in Georgetown with his friends from GW. The spook had suggested a trip to Philly to see a Flyers game and suggested they stay over. Sometime during the afternoon, Mullaney spoke with Paul. He said Paul was surprised and annoyed. He wasn't defensive, but he seemed to consider it a challenge. Nothing more was said for the rest of the day.

On Monday morning, I received a visit from Mike Allen. Mike asked me to cut orders transferring the spook out of the Pentagon. When I asked where, he shrugged and said I could put him at Fort McNair on a temporary duty assignment, until they figured out what to do with him. He said it had to be immediately. I asked why.

"It seems like our man made real progress. He gained the confidence of your man, Paul, and was going to meet the suspected GW drug dealer on Friday night in Georgetown. If things went well, he and your man were going to Philly on Saturday night, where he hoped to meet the connection up there."

"So?"

"It seems our man started drinking to fit in. He got real friendly with Paul and his friends, putting his arms on their shoulders and singing. Someone took out their cigarettes and our man said, 'Give me one of those fags.' He just got back from England where they call cigarettes fags. Well, your man Paul told him none of his friends were fags, and he should keep his hands to himself. Our man said he wanted a fag 'bad.' That's when the fight broke out. They beat the shit out of him. He thinks Paul is the one who took him to the ER."

"So, that's it. The investigation's over?"

"Yeah, I'm afraid so. Looks like your man wasn't involved in any of the drug stuff, and the guy at GW was all smoke, pardon the pun. He talked a big game about the drugs, but he was only trying to impress the ladies. We're back to square one. Someone is dealing to the soldiers, but we don't know who. Thanks for your help."

"No problem."

CHAPTER 34

"Burn and Stir"

Summer was on its way out, but September proved to be just as hot as August. Everyone in the office had returned from vacation or leave, and I settled in for my last eight months in the Army.

I was studying for the MCAT, the Medical College Admission Test, to see if I could get a better score the second time around. I always scored well on the math portion of these tests, but struggled with the verbal section. When I took these things for the first time in college, the difference between math and verbal shocked my advisor. Perfect scores were 800. I got a 790 in math the first time around, and a 520 in verbal. The unimpressive 520 was made worse when he told me I got 200 as a baseline just for signing my name. He humbled me further when he said, "With a verbal score that low, I'm surprised you could read and understand the math questions." His words motivated me as I prepared for my second try at the MCATs. I bought the self-help books and studied them most lunch hours in the JAG library. Every two weeks, I took one of the practice exams, verbal only, to map my progress.

Paul requested time off for Rosh Hashana, observed from Wednesday, September 30, until Friday, October 2. Nobody else requested the time off, so I said okay. While we didn't worry about passes for the weekends, Stoner remained strict about using leave time for days out of the office. He approved the leave for Paul right away. Once I approved his leave for Rosh Hashana, he asked for additional leave for Yom Kippur, October 9. Being a work day, I had to pass it in front of Stoner once I approved it. Stoner came back to Paul with the paperwork.

"I approved Rosh Hashana; now you want another day off the following week, for Yom Kippur?"

"Yes, Sergeant Major," Paul replied with full military bearing.

"Are you going to request time off for Chanukah also?"

"Yes, Sergeant Major. They're all major Jewish holidays, and I would like to spend them with my family."

"When is Chanukah this year?"

"December 22 through December 31."

"Well, Private, I've got a problem. The first is you're asking for three holidays off. You have the leave to cover them, but so do the other soldiers in here." With this, he turned to me. "Do you plan on taking any time off during the Christmas holidays, other than Christmas Day, when the building is closed?"

"Yes. I had duty last year, and I've made plans to visit my parents this year."

"West. One of you has to cover this desk at all times. That means, either you or Paul must be here over Christmas. Are you willing to give up your time, so he can have time off? Before you answer, I want to remind you he is not only the junior enlisted man on your team, but he is the junior enlisted man in this office."

341

"No, Sergeant Major, I'm not prepared to do that. If my plans change, I'd be happy to split the time with him."

Paul sat up straighter in his seat. "Sergeant Major. I'm the only Jew in the office and I have to go on record as saying this might be interpreted as discrimination against me."

Stoner glared at Paul. "Don't you pull any of that lawyer-wanna-be crap on me, soldier. There is no discrimination in this Army of any kind. But there is fairness. There are three religious holidays you want to celebrate, and you already have one. You are the junior enlisted man in this office, and you get what's left over. If West wants two weeks off at Christmas, he goes, you stay. No questions, no discrimination. If you were Christian and West wanted two weeks off at Christmas, he goes, you stay. Religion doesn't make any difference. Rank has its privileges. Period! Comprendo?"

"Yes, Sergeant Major. What about Yom Kippur?"

"West, do you have anything planned for Friday, October 9?"

"No, Sergeant Major."

"Looks like you got two religious holidays, soldier. Don't press your luck."

"Yes, Sergeant Major."

When Stoner left, Paul started in on me, trying to wangle me into committing to time in December. I put my pen down and looked him square in the eyes.

"Paul, consider yourself lucky. If I hadn't picked you out of the clerk school at Fort Dix four months ago, your ass would be in Vietnam right now, with zero holidays home with your folks. Stoner's right. You got two of the three holidays you wanted. If you had asked me about Chanukah, I would have told you the same as Stoner. Last year, as the junior man on Andy's team, and the junior man in this office, I worked Christmas week. I am not giving up time with my family for you." I wanted to end the discussion by telling him I had saved his ass from the FBI and the 902nd with a ruse, but I couldn't.

"But Jonathan…"

"Don't press your luck, Paul."

The discussion was over. I made my point. Unfortunately, my relationship with Paul became cool from that point on. In the big picture of things, it didn't make any difference. He did his job as well as before, and it wasn't like he would be the exception and we would remain friends after we got out of the Army; no one did.

Suchanek called with the news my recruiter had arrived in Vietnam. He hadn't seen him yet, but had arranged to have him report for in-processing. Once again, I reminded Suchanek of my insistence no harm come to him. Phil understood I hated the man for what he did to me, and what he had probably done to others, but agreed no harm would come to him.

Suchanek had a plan. He knew recruiters could be very successful in Vietnam. Most of them met or exceeded their quotas, which meant higher performance scores, which led to promotions and choice assignments when they returned to the world. He also knew a few recruiters sought quicker returns, returns they could spend.

Most soldiers hated Vietnam. Besides the constant heat and humidity, monsoons plagued the country for most of the year, and they lived life in mud. On top of the misery of the bugs, bad food, hard work, and boredom, lay a blanket of terror, as constant and oppressive as the heat and humidity. No place in the country offered safety from attack by the enemy.

Every soldier sent to Vietnam faced twelve months of in-country duty, interrupted by one week of Rest and Recreation (R&R) in Japan, Hawaii, Australia, or any number of places within a reasonably short distance of Vietnam. Girlfriends, wives, families joined the soldier during the week. This one week served as the tonic the Army intended it to be. It became an intermediate target the soldiers aimed for, only six months from when they landed in-country. After R&R, return to the world was only six months away. The tour of duty became reduced to two six-month tours, not one twelve-month incarceration.

But, for others, the one week of R&R served as a breaking point. It was a return to sanity, clean beds, good food, familiar looking women, air conditioning, and all the things they had given up in Vietnam. They didn't want to go back for the remaining six months. These were the fish the recruiters sought, as eagerly as fly fishermen seek trout in a mountain stream.

Recruiters tried to get themselves assigned to the major personnel centers for USARV and MACV. They all had a primary or secondary MOS in administration, so they attached themselves to a major unit and had a dual function. The Army assigned many to a formal recruiting unit in Vietnam. For soldiers re-enlisting, the Army accommodated them. As a result, the recruiter's job in Vietnam was easy. Most recruiters were legitimate, but a few were crooks and con-artists, preying on desperate men who didn't want to go back into the jungle.

More than once, we heard of stories of recruiters who made money selling assignments for re-enlistment. They'd tell a soldier, desperate to get out of Vietnam, no assignments existed in Germany, but the recruiter had a contact who might work something out. It would cost money, not for the recruiter, but for people up the chain. Of course, the money went straight into the pocket of the recruiter, sometimes as much as a soldier's entire re-enlistment bonus.

Suchanek and I had no way of knowing whether my recruiter was a crook or a straight shooter. Based on the way he handled me, we assumed, at the least, he was a guy who made his quota regardless of what he had to do. That was enough. Suchanek decided this guy shouldn't be a recruiter in Vietnam.

Phil told me the recruiter presented himself the following day. He was an E-6 Sergeant while Suchanek was a Spec 4. The Sergeant tried to pull rank on Suchanek from the start. This only made Suchanek's hostility toward the man personal. He said, "Sergeant, you can try to throw your rank around here all you want, but be advised, if I think you're trying to pull rank on me, I will kick this up to my Captain so fast it'll make your head spin. And, the Captain loves to pull rank on Sergeants who pull rank."

The recruiter didn't like being talked to like that, but he had been in the Army long enough to know you don't screw around with the enlisted staff in Personnel. Clerks can lose paperwork and make life miserable. They can put the wrong routing on a pay voucher, so money takes forever to get to you. No, you don't screw around with Personnel.

"Let me take a few minutes to review your file, Sergeant," Suchanek said as he opened the recruiter's file. He had removed the chair from alongside of his desk, so the recruiter had to stand. "You were a recruiter in Albany, New York, before you rotated over here." He paused, thinking. Phil's eyes explored the ceiling, searching for a memory. "Why does your name and Albany sound familiar?" Back to looking at the paperwork, he said, "I wasn't prepared for a recruiter. I was prepared for a 71B40, Senior Admin Supervisor in a line company. Do you want to be a recruiter, or do you want me to assign you to a combat unit?"

Suchanek had asked this question hundreds of times, to hundreds of guys, with different MOSs. The answer was always the same. They didn't want a combat unit. "No, Specialist. I'd just as soon stay here and wait for a recruiter slot."

"Don't know if I have one. I'll have to make some calls, and you can come back tomorrow. In the meantime, I'll assign you to the HQ Company here, and you can supervise some general duty soldiers until I find something for you. I only work nights, so come back tomorrow after you finish work, say about 2200 hours."

"That seems kind of late, especially if I have to put in a full day tomorrow."

"Not to worry, Sergeant. You won't be able to sleep for a week because of the jet lag; twelve hours lag is a lot to overcome. After that, you won't be able to sleep because of the planes taking off and landing all night. The only time there's no aircraft noise is when the VC toss a few mortar rounds in here. Can't have aircraft on the runway then, can we?"

Phil assigned him to the HQ Company for the Army at the airbase. Because of his TDY status, he had no specific responsibilities, but Suchanek had arranged for him to be in charge of the "burn and stir" unit, the group that emptied the latrines. A nasty job made worse by the soldiers assigned to perform the actual work. They were soldiers awaiting a disciplinary hearing, or who had already had their hearing and awaited assignment of punishment, or had been assigned to the "burn and stir" unit as their punishment. Army rules in Vietnam mandated all human waste be destroyed this way, to keep it from being used as a weapon by the Viet Cong.

"Burn and stir" was not an easy job. The latrines were free-standing units made of plywood and two by fours, but instead of single units, two units sat side-by-side sharing a common roof. These multi-unit outhouses had large metal cans instead of a hole in the ground, made by cutting a fifty-five-gallon drum in half. They extracted these cans from the back of each unit through a trap door, hinged at the top. They lifted the trap door, reached in, grabbed the drum, and dragged it out. Before closing the trap door, they inserted another drum, cleaned and sanitized. They hauled the dirty drums to a garbage area, away from any other structures. This not only got the smell as far away from everyone else as possible, but served a safety precaution. Once in the garbage area, they poured diesel fuel or gasoline into the drum and ignited it; the burn cycle. When the fire subsided, they'd use wooden paddles to mix the remaining contents; the stir cycle. They repeated the process until nothing but ash remained. "Burn and stir" operated from dawn until dusk. There were a lot of latrines.

When the recruiter reported at 2200 on the second night, Suchanek looked up from his desk. "I smelled you as soon as you entered the building, like shit and diesel fuel and smoke. You were 'burn and stir.' You can't stay here. Come back tomorrow night and make sure you shower and get that stink off you before you come in here. It's bad enough working here in this heat and stink without you bringing in more."

"But I showered and changed my uniform."

"Not good enough. Do a better job tomorrow night, or don't come in here." Suchanek exaggerated about the smell. In fact, there wasn't much other than the normal Vietnam stink that pervaded everything. The sergeant left the office, sniffing at his uniform.

On the third night, when the Sergeant reported, Suchanek looked up and said, "Better, but I can still smell you." He got up and approached him. Sniffing, he continued, "It smells like you have a fresh uniform on, but I can still smell shit and diesel fuel. I think it's in your hair. If you can't get it out, get a haircut. Most of the guys who do 'burn and stir' shave their heads. It might be in your headgear too. Tomorrow night, leave your headgear outside. 2200 tomorrow night."

On the fourth night when he showed up, Suchanek did his inspection by nose again. The recruiter had shaved his head. Suchanek told him he smelled okay, but he hadn't had time to look into an assignment for him. "I'll do that later tonight. I get off at 0600." Pausing, he did a quick look to the ceiling again, the "how-do-I-recall-a-memory" look, and continued, "But, I can't help feeling I know you. Your name is familiar, I think. It just seems familiar. Were you ever a recruiter in Chicago? Were you a recruiter in 'Nam before? I know the name, but I can't place it. Don't worry, it'll come to me."

"No to both your questions. Albany was my first assignment as a recruiter."

"I'll figure it out."

On the fifth night, Suchanek dropped the bomb. The recruiter showed up again at the appointed hour. Suchanek had returned the chair to the side of his desk and invited the sergeant to sit. He had his file in front of him and was turning the pages.

"You know, I told you your name was familiar, the Albany recruiter thing. Couldn't place it at first; and then early this morning, just before 0400, I was taking a smoke break, getting ready for my nightly call back to the States. I call the Pentagon twice on my shift, once at 2000 which is 0800 back in Washington. They're just getting in." Phil intentionally added unnecessary details as he talked, to add to the anxiety. He continued, "That's when I give them our requirements that came up during the day shift. You know, guys ready to rotate, but don't have assignments yet. Sometimes, if it's important, the guy over there, puts me on hold, and gets an assignment for the guy right away. He's very helpful that way. He's helped a lot of soldiers get close to home. Sometimes, if there are a lot of assignments, or the request is complex, he takes the information down, works on it all day for me. Then I call back at 0400 our time, which is 1600 in Washington, just before he leaves, and I get the assignments. He always expects my call, but sometimes the AUTOVON circuits are crowded and I can't get through. If he doesn't hear from me at 1600, he sticks around. He knows how important it is to take care of the soldiers. We have the same respect for the soldiers we're serving, and we've become friends."

The recruiter was developing little beads of sweat on his forehead and on his upper lip and fidgeted with his hands and squirmed in his seat.

"When I called him this morning, I asked him to remind me of where he had lived when he got drafted. He told me Albany, New York. Once again, he told the story of how he didn't wait for them to draft him, but went to the recruiting office in Albany and enlisted."

Suchanek paused for effect before continuing, "and then he told me about the recruiter who had let him enlist for something that didn't exist."

Suchanek looked up from the file and said, "I know you. You're the recruiter who fucked over West, aren't you?"

The recruiter was fully alert now, sitting up, wiping the sweat from his upper lip. Leaning forward toward Suchanek's space, he said, "I wouldn't put it that way. Look, I talked to West. He's called me two or three times. Instead of being in for three years, he's getting out after two. Everything worked out good."

"Everything worked out good, not because of you, but because of the U.S. Army and the fraudulent enlistment regulations. West was lucky. I've talked about this with my Captain, and he thinks it might take time to find you a recruiting slot. In the meantime, we think you're doing a great job on 'burn and stir.' We appreciate it, and we think you have a future there. Check in regularly to see about the recruiting thing."

With that, the recruiter became the permanent "burn and stir" NCO, the Vietnam equivalent of Andy Griffith's "Permanent Latrine Orderly" in No time For Sergeants. Six months later, when Suchanek rotated back to the states, he was still at the job, and still there two months later when I got out.

Revenge is highly underrated.

In January, Dexter W. told us he and his wife were expecting a baby in August.

CHAPTER 35

"Home on the Range"

With the start of 1971, I had less than four months until discharge. I had taken the MCATs and scored well, and now considered George Washington University for medical school. If I chose GW, I'd be able to stay in the Arlington apartment with rent more reasonable than in the District, but I'd have to commute into the District every day. At best, it would take an hour each way, taking ten hours a week from study time. GW, while they had accepted me, didn't offer any tuition relief or grant money.

January saw dramatic changes in the office. The phased withdrawal of troops from Vietnam was in full swing and being given scrutiny by the press. Teddy left the Army with less fanfare than Andy, and the responsibility for providing the numbers for Mr. Laird fell to Ray. Unlike Teddy, Ray didn't use the random number generator; his numbers were real and meaningful. Discrepancies had to be addressed and explained. While Teddy had been able to provide the numbers at will because he made them up, Ray spent Wednesday and Thursday preparing the numbers and supporting documents.

As the number of soldiers in the office decreased, the civilians tried to get used to working again. Paul had been the last replacement six months ago. Since that time, we had lost Mullaney and Vinnie to discharge. The other offices in the building were losing their soldiers to attrition, as well. Les, Dexter W., and I would leave in April. Only Paul and Ray would remain, as junior enlisted men to support Stoner. Stoner had put in his retirement papers and would leave at the end of the summer.

The NSA group in the building wasn't losing people. NSA, the National Security Agency, had several responsibilities, most of which were above my security level, so remained a mystery. Because I was the 902nd MI liaison for the Pentagon and the Military District of Washington, I made several assignments for them. NSA had a headquarters in Fort Meade, Maryland, about thirty miles north of Washington, but had satellite offices scattered around the Washington area, with one in the Pentagon, called SIGINT, which stood for Signal Intelligence. Part of that group's responsibility involved the protection of the government communication and information systems.

Early last year, I began working with my NSA contact, a real spook, not like the 902nd guys. He'd show up at my desk, unannounced, and give me a folder with very specific instructions. After I completed the requested action, I always had to return the folder. He didn't use the phone to conduct business, or to even call and tell he was on his way. Often, I would return from lunch, coffee, or a break, to be told, "That spooky guy with the mustache was here looking for you." Sometimes an hour, a day, or even a week might go by before he returned. I'd suggest he leave a number or a message, and I'd call him back, but it wasn't his way. He was either here, or he wasn't.

Over the course of the year, I learned little about him. Nothing significant, and nothing to do with his job, not even the location of his office. Selective Service had drafted him, but he enlisted in the Army

352

Signal Corps, going to Fort Monmouth, New Jersey, for communication training. While there, the NSA recruited him.

Before joining the Army, he had been a commercial art student in Boston, as was his wife. They lived in a trendy part of Georgetown, favored more by students than the moneyed politico types. The superintendent at the apartment building had a reputation for keeping the security deposit and the extra month's rent, regardless of how clean the apartment was left. He and his wife made their apartment as comfortable as possible and fitting to their artistic tendencies. What they did to their apartment would also make the landlord earn the money they felt he stole from tenants.

Decorating their apartment, they chose their art not from their collection, but permanent art. White walls served as the canvas for frivolous art during this temporary detour in life. The transformation started in the living room, by painting a black circle on the wall— not just any black circle, but a perfect black circle with an eight-foot diameter. Yes, eight feet, floor to ceiling, in the center of the wall behind their couch. They liked the look but couldn't see it, let alone appreciate it, when seated on the couch, so they painted another large circle on the opposite wall—this time red, again eight feet, floor to ceiling.

NSA filled a lot of positions in 1970, and continued the trend in 1971. This increase in their activity was a direct result of the Pueblo Incident. USS Pueblo was an NSA vessel captured by the North Koreans in January 1968, as it collected radio transmissions off the coast of North Korea. The North Koreans claimed the Pueblo violated the international boundary, and captured the ship and crew. After almost a year of torture, North Korea released them in December 1969. The notoriety of the event put the NSA on the front page of every newspaper and made it a featured story on the evening news for most of the year.

Such notoriety is beneficial in getting funds from Congress, especially when there's an overtone of national security. The NSA spook was loading up empty desks all around Washington. The strategy was to get them while they could.

Knowing he was due to get out about the same time I did, I asked, "Who's going to be doing this job after you leave?"

"No one. Over the next couple of years, all the soldiers will be replaced by civilians."

The advantage of working at the Pentagon, compared to being in the real Army, was we weren't in the real Army—merely working an eight-hour day, five days a week, with no guard duty or KP. We could go anywhere we wanted on weekends, as long as we reported for work on Monday morning. No one remembered taking the Annual PT test or annual qualifying with a rifle.

All that changed when we got a new Commanding Officer for HQ Company in January 1971. The new CO was appalled no one in HQ Company had been required to do any physical training or weapons training in years. It wasn't hard to understand how he found out about the lack of PT. One look around either of the mess halls at Fort Myer and he would see draftee soldiers very much out of shape. Compared to the Third Herd across the street, the HQ Co soldiers looked like pillows tied in the middle with a rope; they were soft and overweight.

The new CO, a Major, came straight from Vietnam, where he had commanded a Ranger Company. An up-and-comer, he considered this assignment another feather in his hat. The problem was he thought he was commanding an elite group of soldiers who were working in the Pentagon, while he only commanded a bunch of college guys wearing soldier suits five days a week. His First Sergeant had been at Fort Myer for a while, having had a run-in with me and Stoner a year ago; so he knew the score and tried to dissuade the Major, but to no avail. The Major scheduled a PT session for all HQ Co personnel on the first Saturday morning in February. The order directed us to report at the Parade Grounds where that first airplane had crashed. Everyone had to report in fatigues with combat boots, or proper Army gym wear.

On that Saturday morning, a cold rain fell, so we moved ourselves to a large maintenance shed near the Parade Ground that barely accommodated us. Most of us had no idea where our fatigues were, and all of us had thrown our combat boots away. We reported in gym wear, some wearing sneakers, and some wearing flip flops or sandals. Wooden clogs were becoming a fashion statement, and a few of the guys showed up wearing them. Most showed up with a coffee, a few with donuts. We milled around, drinking coffee and smoking until they found us. They had waited at the Parade Grounds in the rain.

The Major had a bull horn, and after blasting us for not being at the parade ground, gave a pep talk about how he was going to get us all in shape, make us proud to be soldiers. Then he announced the following weekend he had scheduled the annual rifle qualification for us at Fort Meade, Maryland. After that, we would continue reporting for PT on Saturday mornings, until all of us passed the Army basic physical test. This basic physical test required a soldier to run a mile in less than eight-and-a-half minutes.

The First Sergeant lined us up and made us do calisthenics, then he made us run around the maintenance shed. That proved too crowded, so they led us outside to run in the rain. Except for a few athletes, no one ran the mile in a time close to acceptable. They dismissed us after three hours, and told us to get into a daily exercise program, or the Major would mandate daily exercise sessions every morning at 0530, just like the real Army.

There was strong dislike for this new commander, but there was no plan to oppose him. It seemed he held the cards. We had two weeks before the next scheduled Saturday PT session, but we had to get through the rifle qualifying first.

As the week wore on, the unrest with this situation grew. All of us had gotten into a very nice routine of working a forty-hour week, with weekends free. Many used the free weekend time to get part-time jobs, which were now in jeopardy. But, what to do? During the week, we got revised orders. The Fort Meade rifle range couldn't accommodate all the soldiers from Fort Myers at one time. They split us into two groups to go for rifle qualification on consecutive weekends, and those not at Fort Meade would report to the parade ground for PT.

The following Saturday was overcast and misty as we reported as ordered to the Parade Ground, to be bused to Fort Meade. Once there, we were subjected to more real Army types, who soon became understandably pissed off at us because we were a rag-tag group. Many of us had still not found our fatigues. Of those who had, some had borrowed combat boots, others wore their Army-issued black shoes, and some wore sneakers. Many wore khakis. I still had my old canvas khakis and wore them wrinkled—not as bad as Peter, but not suitable for the Pentagon. I wasn't going to risk ruining my good wash-and-wear khakis.

The way we were dressed was the first thing that got the Fort Meade guys pissed at us. Our First Sergeant was pulled aside and talked to. We saw him shrug his shoulders. The second thing that got them pissed off was the complete lack of military discipline. We didn't line up the way they wanted us to line up. When they finally got us in some sense of order, they divided us into groups, which they called lines. Each line would go to the firing area and shoot as a group. Reacting more to the way we dressed, they forced us to shoot from the prone position, that is, lying down, getting our khakis dirty.

"If you was wearin' fatigues, you wouldn't worry about getting those fancy pants all dirty. Now get down."

The shooting lines were further divided into teams of two. One man would shoot, and the other would spot—that is, he would tell the shooter where the shots hit by looking through a telescope trained on the target. The line not shooting watched the activity from bleachers behind the shooters. It didn't take long for the firing range staff to become unhappy with the shooters. They shouldn't have, though, because the shooters saved them a lot of work. After a group fired, the targets had to be changed for the next group. Only a few of the targets were even hit, so very few targets had to be replaced by the range staff.

When I went to the firing line, they paired me with a guy who had been on the Ohio State rifle team. Interesting, because I had been a competitive shooter in high school. At first, we thought it would be fun to test ourselves with a little competition, but thought better of it after seeing the poor performance of the first line. The rest of the guys would hate us if they used our scores as an example, and forced the others to come back and shoot until they did as well, or at least better than they did the first time. That could make for a long day. Instead, we decided to test ourselves by shooting at things other than the target, like clumps of dirt near the target and twigs. We shot in front of the target, causing puffs of dirt to shoot up, and shot the clips that held the target to the carrier. In the end, our scores proved no better than anyone else.

We thought that was it, but learned the fixed target represented only part of the test. Keeping the same pairings, they sent us through a combat target range. In the combat target range, a shooter would walk along a trail and shoot at man-sized silhouettes that popped up. These pop-ups would be from twenty-five to one hundred yards from the shooter. Neither of us had seen this range before and thought it would be good fun, and agreed to a little competition. Each session would have ten targets, requiring the shooter to reload part way through the range. Ohio State shot first and hit eight of the ten targets, missing one at one hundred yards, due primarily to slow reloading of his rifle. During my turn, I knew of the quickness of the reload target and was prepared, scoring nine of ten. After we turned the scorecards into the range NCO, we took seats in the bleachers.

When the final line finished shooting, the range NCO stood in front of the bleachers with a clipboard. "You guys, I will not dishonor the word 'soldier' by calling you soldiers, you guys are the sorriest group of assholes to ever, I repeat, ever come through this range. Most of you didn't hit the fucking target, let alone score enough points to qualify. You are a disgrace." As he said this, looks of approval passed amongst us. The NCO was right, you couldn't mistake any of us for soldiers.

"They told me you were the smart guys, the guys who work at the Pentagon, the college boys. I can't make you stay here until you get the scores you got when you got out of Basic Training. I can't make you be soldiers." At this, we couldn't hold it any longer. We knew we'd get back on the buses, so we cheered and applauded. This really pissed him off.

"As you were," he shouted. "Two of you are mine. Two of you have falsified official Army records, like you consider this exercise today some kind of joke. Falsifying Army records is not a joke. Two of you are not going home. Two of you are going to meet my MPs and my JAG Officer."

Everyone went quiet. What was he talking about? Where were the records we falsified?

"West and Doane, up front, now."

Doane and I looked at each other with "What the fuck?" expressions. We didn't stand up quick enough for him, so he shouted for us again. Standing, we made our way to him.

"These two guys scored zero on the stationary target, hitting dirt and chopping down targets with their wild shots." The bleachers erupted with cheers and applause. We smiled and took a little bow. "Yet, they scored almost perfect scores on the combat range." The crowd reacted to this news with boos. We both smiled and gave fist pumps, and while I bowed, Doane did a little curtsy. "That is a falsification of records, and for that you will pay."

"Sergeant," Doane said, "The scores were correct. There was no falsification of those records. We couldn't get the sights right on the stationary targets, but we got squared away for the pop-ups. So, we did a little better."

"You did a little better? A little better? You recorded one miss and two misses."

"Yeah, I missed the target that came up while I reloaded. I was too slow. West learned from me and was ready."

"Bullshit, you lied on the record."

"Not so, Sergeant."

The guys in the bleacher stomped their feet and started a chant, "Shoot out! Shoot out!" The NCO looked at them, looked at us.

"To the range, you two. I'll keep score," he said as he led us to the stationary range in front of the bleachers. Giving Doane a rifle, he told him to take the prone position. He called out the targets downrange in yards. "One hundred yards." Doane aimed, fired and hit the target.

"One-fifty yards." Again, same result. Five shots, five hits. When it was my turn, the results were the same. The guys in the bleachers erupted with wild hooting and cheering, with a few throwing hats in the air. The NCO sent us back to the bleachers. Everyone wanted to shake our hands or slap us on the shoulder. The range NCO dismissed us and told everyone to get on the bus.

Our First Sergeant said nothing to anybody, keeping to himself at the front of the bus. On the way back, Doane and I got to talking about shooting, but switched back to the range NCO. Agreeing he was an asshole, but we acknowledged we had misled him, and his response was proper. We also agreed our new CO at HQ Co might be a better fit in the category of asshole.

"If he was an enlisted man, I could make his life miserable," I said, and then related an abbreviated version of what happened with my recruiter.

"The recruiter thinks you did something when you didn't. With the Major, we'd have to make something happen. Different all together," Doane said.

"Besides, the Major just got back from Vietnam, so we couldn't get him on a relocation right now. But, he's such a gung-ho asshole, he'd most likely enjoy going back to Vietnam. We need something else, something personal."

After some thought, Doane added, "I work in the Officer Branch of OPO—not the same job you do, but I know guys who might be able to help."

Doane agreed to do the research, first to determine if he had access to the Major, and second, to think about what he might do. We agreed to talk by phone.

Stoner told me the HQ Co Major was pissed about the showing of his "troopers" at Fort Meade, both military bearing and performance on the range. He felt we reflected poorly on him as the commanding officer, and he wanted his pound of flesh. The First Sergeant cautioned him to back off and let things revert to the way they were before. The CO reminded him that he, like the "troopers," were still in the Army and should behave like Army.

On the Monday after the rifle range trip, I had to remind the Army I was still in the Army. When I put my IBM card under my blotter, I ceased to exist in the Army's eyes. I still had a presence for local issues, like orders from the HQ Co CO, medical at Rader Clinic, and the finance guys, but now, as I got ready to be discharged, I needed the rest of the Army to know. For example, if I decided to go to medical school some place other than GWU, the Army would ship my household goods, if I had household goods.

In the middle of this, Doane called and told me he looked at the

CO's file. In the Officer branch, he had access to the entire file, not just the IBM card I used for the soldiers who were my responsibility.

"I have an idea," I said. "It's really two ideas. The first just sends a message. He might not get it, but his First Sergeant should. If he doesn't get the message, we can follow up with something that should get him more focused."

"What have you got?"

I told him what I was doing for myself to get back on the Army grid. "It occurred to me I might do the same thing for him, and arrange for his household goods to be picked up for shipment. I'd run it through the same civilian outfit that would do any relocation. The movers won't have any idea he's not eligible for a second move. They'll just show up at his quarters and start packing. All it takes is an authorization."

"I like it. When do we do it?"

"I think there would be a certain poetic justice if the packers showed up at his quarters while he's running our asses around the Parade Ground on the Saturday after next."

"It would be nice to avoid that next PT drill, but I do like the poetic justice. Can you see him getting a message from his wife while he's got us running around? He'd have to leave. Let's do it. What if he doesn't roll over?"

"The second one would definitely get his attention. We'll have his pay check sent someplace other than Fort Myer, say Fort Monmouth, or Fort Monroe—an easy mistake, they all start with M."

Doane was impatient. "Let's do both, the paycheck first. The end of the month is next week. If we have to wait until after the movers, it'll be another month. This way, we can pop him twice in the same week. He should get the message."

Doane provided the necessary information, and I made the arrangements. Afterward, through Stoner, we found out everything that happened. The paycheck was missed at once. The family budget was running from paycheck to paycheck, and he had a minor panic, and he asked his First Sergeant to trace things down. When he asked, "How the hell did my check get sent to Fort Monmouth?," the First Sergeant is reported to have said, "I told you not to fuck with these guys, sir."

Later in the week, the moving company called his wife to confirm the packing for Saturday morning. Again, he asked the First Sergeant to straighten things out, and again the First Sergeant reminded him he should return to managing college boys, rather than commanding soldiers. Seeing the wisdom of the experienced First Sergeant, he cancelled PT, as well as the annual PT qualifier. I would have liked to thank the First Sergeant for his help, but that would have left me and Doane wide open for who knows what.

As it turned out, letting him know wasn't necessary. Stoner told me the First Sergeant knew who did it. He asked that his thanks be distributed widely to those responsible. We had made his life easier too.

On the following Monday morning, I held the big oak doors on the Mall Entrance for a soldier struggling with crutches. As he passed, I asked, "'Nam?"

"No, I got mugged in the District Friday night."

CHAPTER 36

"Eli, Eli, Yo

Over the course of the past six months, I spent more and more time with Mike Allen. Part was because of our membership on the joint task force with the FAA and the Air Force, and part because of my assignment to the 902nd MI Group. I didn't know I was also auditioning for a job as a CIA contractor.

Over lunch one day in early February 1971, Mike revealed his plan for me.

"Jonathan, do you remember last week when we went out for pizza and beer after our joint task force meetings?"

"Yeah, why?"

"We were sitting around with some other guys from the meeting talking about possible scenarios regarding increased airline security, ways the government could get everyone on the same page going forward."

"Yeah. I suggested we needed a U.S. hijacking to make people aware of the potential threat. Better yet, an example that raises awareness, without an actual hijacking."

"Right. You suggested a government-sanctioned operation to put the threat on the front page, without putting anyone at risk. While you didn't say it, I think you suggested a clandestine operation to hijack a U.S. plane within continental U.S."

I stopped eating my sandwich and looked at Mike. "I don't recall saying that, but it makes sense. The FAA and the Air Force are at odds about the way forward. The airline industry is opposed to anything like the El Al security checks. Any proposal for such pre-boarding screening would have them marching to every congressman to fight the proposal. At the very least, any proposal would take years to implement, and be a watered-down version of what's needed. From the background presented at the meetings, it's clear terrorists will strike in the U.S. soon, and the results would be as devastating as they've been in Europe. Lives would be lost due to inaction now. If a hijacking occurred here, the public might insist on increased security immediately."

Mike listened with interest as I relayed my recollection of that earlier discussion. "I've mentioned your proposal to my boss, and he'd like to talk to you."

"Your boss in 902nd?"

"No, my real boss. Can you break away for an hour?"

"Sure, I think so." I was short, and with it came a certain leeway toward time in the office. "I'll just tell Stoner I have a meeting with 902nd."

"Great. That won't be an untruth, but it won't exactly be the truth."

Mike had my curiosity up with his statement, and my face reflected it.

"I can't answer any questions now. I'll be in touch."

The next morning, I got a call from Mike asking to meet him on the fourth floor A Ring, at an office overlooking the internal courtyard and the Ground Zero Café, closed for the season. Inside the office, the furnishings were as severe and bleak as the February weather outside. The office looked like someone had just moved in or moved out—impersonal, not even the obligatory picture of the president on the wall.

Besides Mike, there was another man. Prim was the best description I could give him, and medium in every detail: height, weight, and appearance, everything except his shoes and eyes. His steely gray eyes, with emphasis on the steely part, went well with the austerity of the office and the weather. His shoes were distinctive—brown Florsheim shell cordovan loafers, buffed to the shine only a cordovan shoe can achieve. I knew, because I coveted those shoes even more than I coveted my Third Herd patent leather shoes. If he had asked me, I would have traded even up, right then and there. Later I noticed his hands, perfectly manicured.

Mike introduced him as Mr. Anderson; no title, no first name, no other description. Without the offer of a hand shake, Mr. Anderson invited me to sit down at the only chair in front of his desk. Mike took a chair to the side. Mike spoke first, summarizing the hijacking proposal I fabricated the week before. When he finished, Mr. Anderson asked me if I had anything to add.

"It wasn't a proposal, just an attempt to win a free beer. The guys—"

Mr. Anderson put up his hand, the universal sign to stop whatever was going on, in this case talking. "If you have nothing to add, permit me to speak." Without waiting for my permission, he continued, "Your proposal has merit, not much. Mike presented it to my staff at our meeting, and it generated a lot of discussion. They were nearly unanimous in their conclusion: Your proposal was neither feasible nor practical. There was one vote in favor, mine, and my vote is the only one that counts on this."

For the next thirty minutes, Mr. Anderson recited the history of airplane hijackings around the world. Following that, he discussed three different scenarios that played off my original "what if," beer-earning proposition. Mike then asked me to leave the room while he and Mr. Anderson talked.

When I returned, they continued, shifting the discussion to the details of the plan. Curious, I asked why the 902nd was involved in this plan.

Mr. Anderson responded, "I'm not 902nd," and looked to Mike.

Mike acknowledged he was 902nd, "Sort of."

"Something else?" I asked, and started with the alphabet agencies, "FAA, NSA, FBI?" Running out of the domestic government agencies, I added the only other one I knew of, "CIA?"

Both Anderson and Mike acknowledged they were CIA, Mr. Anderson with a blink and Mike by raising two fingers. When I asked about the CIA operating domestically, they shrugged, adding there were exceptions.

By the end of the day, they had reeled me in. I was a CIA independent contractor, with a single assignment of hijacking a domestic airplane, with the help of the CIA. Besides doing my duty as a citizen, they would guarantee my admission to the September 1971 class at Yale Medical School, tuition and board paid, and a substantial tax-free cash stipend. If things went well, they would continue to pay me as a consultant through graduation.

No more studying for the MCATs. Most important, they would teach me how to jump out of an airplane.

CHAPTER 37

"Bad News, Good News"

In early March I got a call from an agitated Dexter W., wanting to meet for lunch. Because it sounded important, I agreed to meet him that day. Still too cold for the Ground Zero in the courtyard, we met at the Navy Mess, a favorite of mine for the steamship side of beef. Dexter W. was at the entrance when I arrived, looking as unhappy as he had sounded on the phone.

"Dex, what's up?"

"Got a problem I have to talk out with you. Let's get our food first before I begin."

We went through the line, both of us opting for the featured beef. Once seated, Dexter W. played with his food as I devoured mine. After the first few bites, I slowed down and gave him my full attention.

"So, how can I help?"

"It's about Kathy, or rather, it's about the baby."

"Is everything okay?" These were nice people. When I met Dexter W., I liked him right away. A big, All-American-type guy from Michigan, sort of a dark-haired version of Les, with Midwest charm replacing the good-ole boy image. When his wife joined him, I liked

369

her even better, petite and blond to his big and dark. They had me over to their apartment regularly for dinner, sometimes Kathy inviting a single girlfriend to join us.

"Kathy had a prenatal check-up at Radar Clinic last week because she noticed a little spotting, and they referred her to DeWitt Army Hospital at Fort Belvoir. We've been trying to get as much care as possible before I get out in April, because I won't have insurance until I start school in September."

Rader Clinic, the medical facility at Fort Myer, was a basic services facility, providing well-patient care and basic emergency room service, until an ambulance took the patient to a hospital. They named the clinic after a World War II Captain who survived the Bataan Death March, took care of soldiers at a Japanese POW camp, but was later killed on a freighter transporting prisoners back to Japan to work as slave laborers. No disrespect intended for Captain Rader, but when they dedicated the clinic in his name, they should have tried to provide medical care beyond that afforded in a POW camp.

"When's the baby due?"

"Not 'til August."

"That's a long time to go bare without insurance."

"But it gets worse. The doctor at Belvoir told her she'd probably lose the baby," he said as his eyes glossed over and tears formed.

"Oh my God, that's terrible. Did he say what the problem was? What can they do about it? When will this happen?"

"That's part of the frustration. While we waited for the appointment, we tried to find out as much as we could about the prenatal clinic. The way it's supposed to work is an OB/GYN doc has a team of three or four nurses and aides who work with him. There are two examination rooms where they prep one woman for examination, while he examines a woman in the other. Each examination is scheduled for ten minutes. The standard procedure calls for him to see five patients, and then a ten-minute break. That's the way it's supposed to work."

Still enjoying my roast beef, I nodded, showing I was still following and understood. The ten-minute break every hour was standard procedure everywhere in the Army except in Vietnam. Apparently, the Viet Cong wouldn't cooperate.

"This doc carved out five examination areas, the two rooms they allotted him, but then he found three beds in a clinic area, one of those areas with curtains on rails running from the ceiling. With these five rooms, he changed his procedure. He had the nurses and aides prep five women, while he drank coffee and played grab-ass in the doctor's lounge. When all the women were ready, he put his coffee down and examined the women, spending two minutes with each. In ten minutes, he examined the five women and returned to the doctor's lounge to a still-warm coffee for a fifty-minute break."

Dexter W. was becoming visibly agitated, so I told him to take a minute before he continued. His agitation was infectious, causing me to put my knife and fork down and give him my full attention.

"When Kathy told me how the doc told her the bad news, I became furious. After he did a superficial examination, he confirmed she was bleeding and started for the door. She asked him what would happen to the baby, and over his shoulder with no emotion as he left the room, he told her she'd lose the baby."

"Did he say when? Would it pose a threat to Kathy's life?"

"No to both. We've checked with some people, and they say it could be anytime, even up to the time of birth. If so, the delivery might be complicated, and expensive."

Dexter W. stopped talking and took a deep breath. The tears that had been building while he spoke started running down his cheek. He grabbed a paper napkin to wipe them.

"I can't imagine how you feel. How can I help?"

While it had only been two days since the visit to the prenatal clinic at Belvoir, they had gathered a lot of information and given their options a lot of thought. He explained how lots of soldiers found themselves in this insurance dilemma. There were two options, three if you counted going bare. A post-Army insurance program called CHAMPUS was available for situations like this. While it sounded great, it wasn't much more than a pay-as-you-go installment plan costing the same as a normal civilian delivery. That seemed unfair, but it was insurance, not a hand-out.

The other option was for the soldier to extend his time in service to cover the delivery. The only problem was, they only granted service extensions in three-month increments. With an end of April discharge, a three-month extension would only take them to the end of July. With the baby due in August, close didn't count, except for providing health coverage in the event of early complications. Extending another three months until the end of October would cover the delivery date, but was a no-go if Dexter W. wanted to start graduate school in September. They only had a few months to work something out.

Kathy called Radar Clinic for a follow-up, but it was equally disappointing, as they only confirmed she was spotting. With the clinic being the on-site medical facility for Fort Myer, and the clinic for all the

Generals stationed there, you'd think they were more skilled. Maybe they were for the ailments of Generals, and pregnancy wasn't on that list of preferred conditions for General officers and their families.

"Dex, I don't know what to tell you. This isn't my field. I can ask around, but I don't think I can get any better answers than you have."

"We'll continue to look for answers on the medical issues, but that's not the reason I came to you for help. It's the extension I need help on. Is there any way I can get a four-month extension? Based on what I've heard you've done, it sounds like if there's any office in the Pentagon that could get a waiver, it'd be your office."

"I'll ask as soon as I get back."

We spent the rest of the meal rehashing the stuff we'd already gone over, with much the same result of nothingness.

Back in the office, I stopped at Stoner's desk. After I explained the situation to him, he scratched the side of his nose. While he could pull some strings, he admitted this sounded like a policy issue and suggested I talk to the Major. The Major didn't have an immediate answer, either, but agreed to look into it.

That night, as I was going into my apartment, the guy downstairs was returning from putting out the garbage. He was a second-year resident at George Washington Hospital, and an Army veteran. We'd only talked a few times, but I took this opportunity to ask his opinion.

"I rarely get on a soapbox," he said, "but this situation pisses me off. I was a Navy corpsman in the early days of Vietnam, assigned to a Marine unit. Saw some fantastic medicine performed by dedicated doctors. The experience got me interested in medical school. The residency has opened my eyes, though, about how the government treats the veterans and the military families. You'd think in the District area, with all the military here, there'd be top-notch health care. There is, if you're important and can get into Bethesda Naval or Walter Reed. That's not the case for a lot of dependents. We get a lot of them showing up at GW when they should be at a military hospital. When we ask, they tell us about long wait times in the ER, or even longer waiting times to get an appointment with a specialist. Rather than put their health at risk, they pony up the money and come to our emergency room, or see our specialists."

"I didn't realize it was a problem," then realizing he was agitated, I apologized for bringing it up.

"You'd think with all the military here, they'd have a better system, but I guess it's all the military in the area who are overpowering a system not geared to handle it. There are great military docs at Bethesda Naval and Walter Reed. They're the same kind of guys I saw in Vietnam, same dedication, same care for patients. It's not their fault, it's the system. This guy at Belvoir sounds like the exception, but once he signs off on the case, it's tough to move it through the system. Rules and regulations for the majority don't address the exception."

I thanked him, for what, I don't know. All he'd done was tell me the problem was systemic, and he's seen it before.

Two hours later, my doorbell rang, and it was the doc from downstairs. "This thing with your friend bothered me, so I called my chief resident. He's a veteran, and has the same concerns about the medical treatment dependents and vets get. He made a quick call to a vet he knows in the OB/GYN at GU, and they want to see her. Have her call this number tomorrow and ask for this doctor. He knows all the details you gave me, and will set up an appointment for her." Before

I could say any more than thank you, he turned and went down the stairs to his apartment.

I called Dexter W. at home and relayed the information. They were grateful and hopeful at the promise of getting some answers.

Two days later, Dexter W showed up at my desk, smiling. I stood, and he hugged me, tears streaming down his face.

"We just got back from seeing the doctors at GW, and I can't thank you enough."

"Sit down," I said, offering him a chair.

"It was only a polyp. They removed it and said all is well. The baby will be fine. Thank you, and thank all the people who helped us. Kathy says if the baby is a boy, we'll name him after you, and we want you to be the godfather, regardless of a boy or girl."

I hadn't done anything, but these nice folks wanted me to be the godfather of their yet unborn baby. "I'd be happy to do it."

"Kathy's waiting at home. The office told me to take the rest of the day off, so we're going to celebrate by going out to a late lunch. A date, a happy date, something we couldn't have imagined last week."

Dexter W. left, passing the Major and Stoner's desks on the way out. As my eyes followed him, they stopped at Stoner, who motioned me to his desk. As I walked over, the Major stood and joined me there.

"What was that all about?" asked Stoner.

I told them an abbreviated version.

Stoner was the first to react saying, "I've been in the Army for over thirty years, and for the most part, I couldn't ask for better friends or more competent individuals. Do you know why? Don't answer, I'm going to tell you. Because we're brothers; we count on each other, covering each other's back. But every once in a while you run across an asshole like this guy, who puts himself before or above everyone else. And you know what happens? Everyone remembers the asshole. Gives the service a bad name. Makes me sick. You did the right thing, West. You remembered this young soldier is your brother, and you helped him out. Maybe your kindness will erase the sting the doc at Fort Belvoir put in their lives."

The Major had listened to Stoner, nodding his head in agreement, then added, "I've not been in the Army as long as the Sergeant Major, but I agree with him. Not only does it give the Army a bad image, but if he did that to your friend, he probably does this to other patients, some of whom may have serious medical conditions needing more medical attention than he's providing."

The Major's statement hung in the air as we looked at each other, until he broke the silence. "Kind of like what that recruiter did to you. Too bad they don't put officers on burn-and-stir duty."

The Major returned to his desk after dropping that bomb on me. Looking at the Sergeant Major, my eyes still popped, I asked, "What does he know about the recruiter?"

Stoner fumbled with the papers on his desk before responding, "Everything. Everyone knows about the recruiter. Recruiting Command was going to look into disciplining your man, but found out what you did and felt they couldn't do any better. The story is all over the building, and all the recruiters in the country are getting it as gossip through their grapevine. The message has been received loud and clear, to do the job the way it's supposed to be done. No shortcuts, no lies, no promises, or the next guy gets roasted. Fuck up, and you get to be the burn-and-stirrer, if you get my drift."

I returned to my desk and had to let that settle in. A personal grudge had taken on a much broader significance. The system had been bigger and more powerful than me. More disturbing, it showed itself to be uncaring, as uncaring as the career soldiers and bureaucrats who waged war forty hours a week at the Pentagon, and lived normal lives the remaining 120 hours a week, with no regard for the people walking in the mud at the side of the road in Vietnam. But I had beaten it and liked the outcome.

I wasn't comfortable with everyone knowing what I did.

As I thought about it, it wasn't just the doctor at Fort Belvoir. The staff who attended to him were complicit in his actions. No one stood up to his two-minute examination, or demanded he answer Kathy's question more fully. No one displayed any sympathy. Unlike the recruiter, this was not a single individual doing wrong; this was a hospital allowing an individual to do wrong. Like with the recruiter, I had to send a message.

While I had helped Dexter W. and Kathy with the immediate problem, they stilled faced the problems of extension of service and the medical insurance coverage. While I thought about that problem, Dexter W. called me.

"Great news. Kathy's doctor just called again. The hospital and the staff are really stepping up to the plate on this. They've offered her a job at the hospital, part-time, until the baby's born, as a receptionist in the mornings at the OB/GYN clinic. On top of that, they've offered me a job in the accounting department in April, beginning the day I get out of the Army. It's not a gift, they need me to cover for people on summer vacation. Because I'll be an employee, we'll have medical coverage. Jonathan, we're so grateful to you for making this happen. I know you'll say you had nothing to do with this latest chain of events, but you got it started, and that's important."

After hanging up the phone, I sat thinking for a minute. While glad for Dex and Kathy, I was embarrassed civilians had to help out. At the same time, I was proud of the medical community and the people who reached out to help, making me proud I chose medicine as a career.

The revenge against the recruiter took months to fulfill. No so with DeWitt Army Hospital. Like the recruiter, I didn't wish any of the personnel at DeWitt Army Hospital to be put in harm's way. I wanted them to suffer some degree of grief and anxiety.

I knew at once what I had to do to satisfy the rage burning inside me.

CHAPTER 38

"Fiddle Fucking Around"

Throughout the week, my need for revenge waxed and waned, waxing on even days, waning on odd days—or was it vice versa? Only when I focused on the doctor at Belvoir, his immediate team, and the dozens, maybe hundreds of other women he may have treated the same way, did my resolve firm. This action I planned for the maximum exposure. If the news of my scheme with the recruiter had spread and resulted in change, I sought no less for this action.

While our office dealt with individual soldiers being placed against individual assignments, there was an emergency plan where we could activate entire units. While intended for a national emergency, the Army had already used it several times during the Vietnam War to call up Army Reserve Units or National Guard Units. I hadn't done it, but I knew the drill, and where we kept the unit IBM cards.

Stoner took Friday off, which proved ideal. Since the recruiter thing, his eyes never left me when I moved around the office. Sometimes during the ride home he'd ask if the avenging angel struck today. The first couple of times, I played dumb to the reference, but then it became a game.

ത

With the Sergeant Major out of the office, I moved freely. Occasionally, I saw the Major following my moves a little more closely than normal, but dismissed it as being over cautious. In the special unit designation folder, at the back of the top drawer in the SECRET cabinet, I pulled the card for DeWitt Army Hospital at Fort Belvoir. Next, I pulled the general MOS cards for all 7, 8, 9, and zero series specialties. These cards told the computer in the basement to pull all the personnel at DeWitt Army Hospital with these MOSs for Staff Sergeant and below. This included all clerks, supply people, medical staff, cooks, and MPs. Next, I pulled the general assignment card. They would use this card to fill in the "Report to" line on everyone's orders.

I chose MACV HQ in Vietnam. Suchanek enjoyed the revenge against the recruiter, so he'd appreciate this move too. Of course, he'd just watch it and be amused; nothing would happen, because no one at DeWitt would see the orders until Monday morning. When the shit hit the fan, I'd have to issue a revocation, but not before anxiety had filled the hearts and hallways of the hospital for an hour or so.

Finally, I pulled my authority IBM card identifying "Per Spec 5 West, EPCMR-GS, OPO" as the originator of the orders. So what? I was short, less than sixty days!

Before I left for the day, I put the cards in the bin across the hall, and set off for a weekend in Fredericksburg, Virginia, to begin my training for Mike Allen and the CIA. For the next two days I'd be out of touch as the IBM cards percolated through the system, before the turmoil exploded Monday morning at DeWitt Army Hospital.

The plan was flawless, almost. The computer in the basement of the Pentagon would cut the orders on Friday night, to be sent by Teletype to DeWitt Army Hospital in an overnight transmission. Unfortunately, all hell didn't wait until Monday morning to break loose. Because it was an Army post, they worked different hours than the Pentagon. Someone worked in the DeWitt Army Hospital personnel office on Saturday morning. Hell broke loose forty-eight hours earlier than I expected.

When the activation order came in, the personnel clerk passed it up the chain of command. Because of the nature of the order, the command structure wanted to verify the order. Fort Belvoir Personnel responded with a telex to the order, asking Spec 5 West to verify. Specialist West couldn't verify, because he was learning how to dye his hair in Fredericksburg, Virginia. When they didn't get an answer to their Teletype request, they called Spec 5 West at EPCMR-GS OPO. No one answered the phone, because no one works in the Pentagon on Saturday or Sunday. Normally, nothing would happen until they could verify the order. In most cases, personnel involved wouldn't even know, until they verified the order. But DeWitt Army Hospital didn't exactly abide by all the Army rules. Someone in DeWitt Personnel told someone outside of Personnel, and the word spread fast, first throughout the hospital, then all over Fort Belvoir.

In an unforeseen complication of activating DeWitt, I also activated Rader Clinic personnel, because of its status as a satellite of DeWitt Hospital. So, lots of people ran around Fort Belvoir and Fort Myer that weekend, thinking they had orders to Vietnam.

For a unit move using the emergency procedure, the destination got notified as soon as the orders left the Pentagon, alerting them about the arrival of so many soldiers. Suchanek got the orders when he came into work on Saturday night. He told me later he wondered what they had done to piss me off so much.

On Monday, with the Sergeant Major still on leave, I didn't pick him up. When I came into the office, the Major called me before I had taken two steps into the room. He held a copy of the orders, waving them, and asking me if I knew anything about them. I took the orders from his outstretched hand as I followed his motion for me to move behind his desk and take a seat. I figured no one would know about this until later in the morning. The folks at De Witt must start work a little earlier than we did, but I guessed wrong.

As I searched for an answer, he said, "These orders were issued Friday afternoon, Per Spec 5 West, and arrived at DeWitt Saturday morning. There's been a shit storm down there all weekend. They've been trying to get you to verify. Did you do this, or did someone take your name in vain?"

"Taking your name in vain" was a term used to describe an unapproved assignment. It occurred rarely, and usually involved someone in a Personnel section somewhere in the world knowing how we operated. They could move someone and use the authority line of "Per So and So, EPCMR-GS," and make up an OPO Control Number making it look legitimate. We moved so many people, we never found out about it, unless someone complained. The soldiers being moved got what they wanted, so they never complained. Every time they repeated this illegal activity, the odds of them being detected increased. It was a serious offense and kicked off an investigation when we suspected it.

"No, sir. No one took my name in vain." The Major had me, I had no way out except to man up. I reminded him about Kathy's trip to the OB/GYN clinic at DeWitt. My insides were in knots. As the story unwound, the Major put his elbow on the arm of his chair and slumped. His chin eventually found the palm of his hand and came to rest, moving only once. When the Colonel appeared, he lifted his head and waved him off in a manner that said, "No, it's okay. I've got it." Then he returned his chin to his hand. I thought I detected a smile, reflected only in his eyes, but I wasn't sure. The hint of smile gave me hope.

He had only asked if I issued the orders, and I told him everything up to Friday afternoon, stopping at the point of his question.

"Well, did you issue the orders?"

"I guess I did, sir. Probably not thinking too clearly. I was thinking about my weekend, and had some things on my desk to finish before I left. Also, I might have been distracted by what you said, about how other women may not have gotten the treatment they needed. Instead of doing the work I was supposed to do, I accidently did something else."

"Like you accidently put that recruiter in Vietnam on permanent burn-and-stir, and accidently had the movers show up at the HQ Company CO's house?"

"Sort of, sir."

Now he was smiling, not really smiling, more of a conspiratorial smirk. "I know a lot more than you think. Got to. I've never worked with a cleverer group of men than the enlisted men in this office. You guys keep me on my toes. I do everything I can to stay ahead of you. For the record, if it was me, I would have kept the recruiter from recruiting, but I wouldn't have been so clever to come up with permanent burn-and-stir."

"I had some help on that assignment, sir."

"Yeah, I know that too. You and Suchanek are a legend over there. The recruiter is lucky. Someone else would send him to a line unit, but I understand you insisted on no harm. I suspect you knew no one at DeWitt would go to Vietnam. You just wanted to screw with their heads for a while."

"Yes, sir."

"While officially I can't condone it, privately, it was very clever and appropriate. This may help more than just satisfy your lust for vengeance. Medical treatment down at DeWitt might improve, and that's good. The Colonel is upset, but he'll be okay. You'll have to square it with the Sergeant Major, though. Ruined his weekend. Now, we've got to revoke the orders. I already told them down there it was probably a computer glitch. I'll have Ray draft up a revocation order."

"Thank you, sir."

"Let me ask you. How did you figure out how to do that?"

"Andy and I were trying to come up with ways to minimize the extra work Teddy was making for us, and we stumbled across the unit activation procedure. Knowing that, all I had to do was find the cards."

"Could you do that for anyone?" he asked with a smile, like he had someone in mind.

"Yes, sir. Since they did away with service numbers and replaced them with Social Security account numbers, it's easy. If I knew President Nixon's Social Security number, I could put him on orders to Vietnam."

"Do you have my Social Security number?"

"Yes, sir."

"Are we square?"

"Square as square can be, sir. Next to me, you're the safest guy in the office."

He smiled. "Back to work."

Later in the morning, I got called back to the Major's desk.

"Seems I can't be the authority for the revocation. They set the unit activation system up to run in case of a national emergency, and the originator must make any changes. The activation came with your authority line, so, you've got to issue the revocation. To avoid any confusion, revoke my revocation as well."

"Sir. Are you saying my revocation has to include a revocation of your revocation?"

He paused, and looked at me, puzzled, probably not because of what I asked, but perhaps with what I was up to. I wasn't up to anything until he looked at me that way, then I was up to something. I issued the revocation of the revocation, but I worded it in such a way to make the status of the personnel at the Rader Clinic unclear. DeWitt came back and asked for clarification.

The Major called me up again. "The revocation of the revocation has some unclear wording. It seems to leave a question about Rader Clinic. Do it over."

"Okay, sir. I'm not clear. Do I have to issue a clarification of the revocation of the revocation of the revocation, or just another revocation?"

"Make it clear and final that no one at DeWitt or Rader is on orders to Vietnam. No more fiddle fucking around."

"But, sir. I can't confirm that. There may be personnel on orders to Vietnam. Someone may have been called up on the normal monthly pull. I'd have to check on that."

"Not this month. This month, no one from DeWitt or Rader will go to Vietnam." The smile disappeared. The game was over. I had played all fifty-two cards in my deck, plus the jokers. "Is that clear?"

"Yes sir, clear."

As I walked away, he called after me. "Hey. Are we still square?"

I turned and smiled at him. I knew what was crossing his mind. "Yes sir. We're still square." He nodded an affirmative.

Stoner returned from leave on Tuesday. When I picked him up, it was clear he knew what happened. The temperature in the car dropped ten degrees when he got in with one cup of coffee.

"Before you start with the pleasantries, asking me about my weekend, don't. It wasn't nice. Going on leave required I provide contact information. I stayed home. The phone started ringing Saturday about noon. All morning, Fort Belvoir tried to reach someone in the office. Not being able to get anyone, they called the Pentagon Duty Officer, who called the Colonel, who knew nothing, so he called me after trying to reach you at home. After I tried telephoning you, Mrs. Stoner drove me to your apartment, and you weren't there. No one knew where you were or how to get hold of you. The same drill happened on Sunday. Only on Monday, when you showed up at work, were they able to get some answers."

Stoner stopped talking, took a pull on his coffee—only his coffee today, none for me—and continued. "Yes, it was wrong what they did to your friend and his wife; but never pull a stunt like that again." Pausing again, he took a deep breath and added, "Without telling me. Got it?"

"Yes, Sergeant Major."

"The only reason I'm not taking disciplinary action against you is because the Major told me not to. There, I've said what I'm required to say as the NCOIC."

"Thanks. The Major already reamed me out. Told me not to fiddle fuck around anymore. I said 'Yes sir,' but I don't know what fiddle fuck means."

Stoner turned in his seat to face me. "It means don't fiddle fuck around anymore."

And that was it. As far as he was concerned, the issue was closed.

When the final revocation went out, I thought that would be the end of it. Dexter W. and his wife would have nothing more to do with DeWitt, their score settled; I had been chewed out; and everything returned to abnormal, as Peter would say. And, it would have—except for the second-year GW resident downstairs, grilling fresh blue fish.

He had gone fishing, caught a large blue fish, and invited me to join his family for a meal. I drank his beer while he cooked his fish. During our beer drinking and conversing, he stared at me. Stepping away from the fish, he stood in front of me, looking at my right eye. He asked, "How long have you had that?"

"Had what?"

"That growth in the lateral margin of your eye."

"Oh, that thing. A couple of years."

"Has anyone ever looked at it?"

"No. Why?"

"It could be serious. You should have it looked at."

Then he proceeded to look at it for the next two hours, while he finished cooking the blue fish, while we ate the blue fish, and as we cleaned up afterwards. Of course, his wife saw him looking and asked what he was looking at. After he answered, I had two people staring at it. The intensity of their glare diminished only when I agreed to have it looked at.

The next day, I made an appointment at the Rader Clinic. When I reported to the Rader Clinic, they confirmed I had something in the lateral margin of my right eye. Once again, as they did with Kathy's spotting, confirming something known and doing nothing but refer me to a specialist at DeWitt Army Hospital.

When they said I had to go to DeWitt, I experienced none of the earlier rage. In fact, it didn't even cross my mind. My issue had been with the OB/GYN service, and my appointment was with the eye folks, an entirely different part of the body, almost the other end.

Dexter W. told me the wait at DeWitt had been long, so I brought a book and grabbed a coffee before I reported to the front desk. I found an uncomfortable plastic molded chair in one of the quieter corners of the waiting room and settled in. I wore my lightweight green uniform, and unbuttoned the front of the jacket. The room was warm from the mass of humanity crowded into the small space, and late-winter sun shining through the large glass windows. I had just opened my book when a shadow standing between me and the window blocked my light. I looked up to see a Master Sergeant standing in front of me, holding a file with my name on it, with a big red star.

"Are you Specialist West?" I hoped he wasn't the eye specialist! My sleeves had the rank insignia of a Spec 5, clear and prominent. On my chest, which he was looking at, sat my nametag, which said "West," all in capital letters.

"Yes, Sergeant."

"Do you work in the Pentagon?" he said as he looked at the very distinctive pentagonal shoulder patch reserved for soldiers who serve at the Pentagon. There seemed to be a pattern developing for DeWitt Army Hospital personnel. They confirmed pregnant women to be pregnant. They confirmed women who said they were bleeding from the vagina were, in fact, bleeding from the vagina. Soldiers who wore a Spec5 insignia on their arm, wore the nametag "West," and had the Pentagon patch, were most likely Spec5 West from the Pentagon. I passed on the opportunity to comment on this man's power of deductive reasoning. After all, he might hold the key to how long I waited for my appointment with the eye specialist.

"Yes, I do."

"Is there more than one Spec 5 West who works at the Pentagon?" he asked as he looked at my Third Herd shoes, with admiration and awe. Finally, a legitimate question not restating the obvious.

"No, I'm the only Spec 5 West working in the Pentagon."

With this information, he asked me to "please" take a seat, and he would be right back. The Master Sergeant returned to the counter and conferred with several anxious-looking people, also dressed in white uniforms. These were more Army medical people, or ice cream vendors. He left them at the counter and hurried to the back of the

389

waiting room and entered a room. I could see him in a treatment room. There a man stood, looking like he was doing doctor-type stuff to another man sitting in a chair you might see in a dentist's office. They conferred, their dialog accompanied by hand and arm gesticulations, some pointing toward me, then to my file. The doctor-type guy pointed to the man in the chair, making the Master Sergeant more animated. When the man in the chair got up, the Master Sergeant left the office and returned. He motioned me to join him.

I followed him back into the office, arriving at the office door just as the guy who had been sitting in the chair left. He carried his jacket, which had the rank of a full Colonel on the shoulder. I watched him move to the waiting area, where he took a seat.

The guy doing the doctor stuff turned out to be a real doctor, a Major, and the eye specialist I was scheduled to see. He told me to take a seat in the dentist chair. The Master Sergeant hovered, after taking my uniform jacket and draping it over his arm. The doc told me it was unusual for a treatment of one patient to be stopped in favor of another patient, but the Master Sergeant insisted. What he meant was, Why was a Spec 5 being treated preferentially over a Colonel? The Master Sergeant smiled and stroked my jacket. The doctor completed his examination and described the growth as a chalazion, a cyst. Because I had it for a long time, it wasn't something I should worry about. He asked if I wanted him to remove it.

"Is it a problem if I leave it alone? It's been that way for a while. It doesn't bother me."

"No problem to leave it alone. If it bothers you, it's a simple procedure to remove it. Just monitor it. If it becomes inflamed, see an eye doctor."

I thanked him, as the Master Sergeant helped me get back into my uniform jacket. He was so attentive, I expected him to pull a little brush out of his pocket and whisk my shoulders. He didn't.

"Please accompany me back to the front desk to take care of some paperwork?"

When we got back to the desk, a Captain joined us. The Captain asked if I was satisfied with the treatment I received, and if I had any questions they hadn't answered. I told him I was satisfied and thanked them both. They seemed uncomfortable, and soon I learned the reason. The Captain asked if I was sure. Before I could say yes again, he added, "So, you're satisfied with the treatment you received here today, and you won't put DeWitt Army Hospital personnel on orders to Vietnam this weekend?"

"Yes sir," I replied. I shouldn't have paused so long before I said, "I'm happy with the treatment I got here. I'm not aware of any pending orders to activate DeWitt Army Hospital."

Getting special treatment was nice, but it would have been better if everyone received that kind of treatment all the time.

Later in the month, I had some dental work done. Might as well have the Army do it rather than pay a civilian dentist. I received similar special treatment at the dental clinic. If I stayed in the Army, I would probably never have to bring a book to a doctor's waiting room ever again.

CHAPTER 39

"Bombings and Burgers"

March started off with a bang. Literally.

Arriving at the Pentagon on the morning of March 1, 1971, I watched people being searched. Briefcases, women's bags, and anything bulky were looked at with increased scrutiny. The Pentagon has "Embassy-level security," which means access to the building is controlled, and the immediate surrounding area monitored. Normally, I just showed my ID as I passed, single file, between two security officers. Today's increased security came about from the early morning bombing of the U.S. Capitol Building. Instead of shutting down all official buildings for the day, the government ramped up security.

Mike Allen and I met for lunch, a weekly occurrence since I joined the "company."

"Interesting morning," he said, as we started through the line in the Navy cafeteria.

"Yeah, I was surprised to see the searches when I came in, until I learned why. Seemed superficial considering they bombed the Capitol."

"Not really. They were looking for a bomb that would do significant damage. The Weather Underground claimed responsibility, and they're known for making big statements with their bombs. Big statements need big bombs."

"I suppose you're right."

"What I found more interesting, considering our concerns about airline security, was the reaction of the people at the inspection lines. When we heard the government buildings would be open today with increased searches, we sent teams across the city to watch behavior."

"Ah, see how they'd react to waiting in line."

"The results are interesting, and support our earlier thinking about airplane security. Here we have people who are only going to work, complaining about being delayed. These same people use a light dusting of snow as an excuse to stay home because it's too dangerous. Even at the Capitol building itself, the staff, literally within sight of the bomb damage, complained about the delay."

"One person in front of me in line complained out loud, but others fidgeted a lot. Maybe they figured they made a mistake, and should have used the bombing as an excuse to stay home."

"Don't know. One thing for sure, if we extrapolate the complaints we heard from workers to people trying to catch a plane, I'm confident we made the right decision with our plan."

When we settled in for our lunch, Mike told me Mr. Anderson was not pleased with what I did with Fort Belvoir.

"While he agrees and sees the righteous indignation satisfied by the retribution, he doesn't like the attention you've drawn to yourself. In the future, in fact from this moment on, you are to be invisible, or you're worthless to us. Within the definition of worthless is termination of your admission to Yale and your compensation agreement."

"Understood. Won't happen again."

The Major called me to his desk one morning and explained that he had to assemble projections for next year's PCS moves. Permanent Change of Station moves are one item contributing to the overall budget of a unit. In the biggest picture, the unit is the Defense Department, but as it goes down the chain, it includes our Branch. The more moves, the more money. No one wants to be on the short end of the budget stick, and no one wants to be so greedy the budget gets knocked down. The best way is to find out what others are projecting and be in the same neighborhood. In this case, the Major wanted me to find out about the projections of the branch in the next office, the Combat Arms Branch.

"You want me to spy on the CA branch and get classified information?"

"It's not spying. The information will become available in the budget."

"Just not now?"

"Right."

That night, I told Les, and asked him if he was involved in the PCS projection for CA.

"Yeah. The Branch Chief asked me to find out about the General Support estimates."

"I can't get those until I get an estimate of CA projections."

"Seems a simple solution, little buddy. You make up yours, and I'll make up mine, then we'll share."

So, that's what we did. We made the numbers up, shared them with each other, then with our bosses. Two insights here. Within the Army, they gathered information about other units for budget purposes. Extending that logic, it made sense there were people in the Army trying to get similar information about the Navy. The other thing I learned was a reinforcement of what I had learned watching Teddy prepare the weekly Vietnam troop strength for the SecDef briefing: Don't put any credence in government numbers. In my experience, two out of two were made up.

In late March, the government found Lieutenant William Calley guilty of premeditated murder of civilians in My Lai. Of the fourteen charged with the crime, only Calley was found guilty and sentenced to life in prison. All the soldiers in the office were taking about it, all disappointed Calley was the only one found guilty. When I asked the Sergeant Major about it, he waved me off, saying he had done his talking.

"It was a big part of my life for a year, in Vietnam and when I returned. I haven't spoken about it since they made the charges against the fourteen soldiers, and I won't start now.

April arrived with all the promise spring in Washington can bring. Forsythia and green grass led the way, bringing color back to this gray-and-white city of monuments. Cherry blossoms would soon celebrate the official arrival of spring.

Inside the five-acre courtyard of the Pentagon, people would once again flock to the "Ground Zero Café," officially defined as a hot dog stand. I enjoyed many meals in the courtyard, but never a hot dog. The burgers were fresh, charbroiled on an eight-foot grill. The grill had a mechanical mesh top that fed raw burgers in one end, and perfect medium burgers came out the other. So what if the Russians always had two nukes aimed at it? The owl on the roof kept a watchful eye.

Les and I gave notice on our apartment. As he prepared for his return to the life he left, I remained vague about what I would do until I returned to school in September, making references to traveling or visiting my parents. Les didn't press me for details.

The week before we were to get out, the Vietnam Veterans Against the War held a rally in Washington. Thousands showed up, and the local news bombarded us with coverage. The anti-war sentiment was still high, perhaps growing. One vet testified before the Senate, a former officer from Massachusetts. His testimony denigrated the soldiers who served in Vietnam and the government that sent them there, claiming we had interfered in a civil war and slaughtered thousands in the name of democracy. He might have been more believable had he not worn his fatigues and gotten a haircut. As such, he was just another hippie protestor. I'll bet his parents are proud of their baby boy. He's happy. He got his fifteen minutes of fame before slithering back into anonymity for the rest of his life.

Monday, April 26, 1971 was my last day in the office; tomorrow was my wake-up.

CHAPTER 40

"Adieu"

My last day in the office. I had been on a rigid five-days-a-week schedule for the first time in my short life. As a student, classes and lab work dictated my schedule, but no one cared if I showed up. Only the final grade mattered. If I passed, there were no sins. Life as a hockey player had some rules, but not a routine. The last eighteen months introduced me to structure, and I got used to it, but I didn't know if I liked it. Now, I didn't know if I would miss it.

During these months, I grew, challenged with something I had no training or previous experience in. I found I could think, I could work with people, and I showed compassion for others. Most illuminating was learning I could solve complex problems outside my chosen field of science. With science, at least the way I learned it, there were no "maybes" or "almost." If two plus two didn't come out four, then I did something wrong. The trick in science was finding out what mistakes you made, and to get as close to the truth as possible.

The Pentagon taught me the rest of the world was not as perfect. Andy showed me it wasn't a question of "close enough for government work," a cynical outlook, but rather, "that's about as good as you can get in an imperfect world," a practical outlook.

One thing I would miss was the ritual of arrival each day. That first evening, when I wandered through the tunnel after dinner and stood alongside the Pentagon, had a profound and lasting effect on me. The impact of the building, its size and the impact of the activity within its walls on the world stage, never left me. Each morning, as I emerged from the tunnel, it struck me with the same sense of awe. As time passed, and I became part of the function of the building, albeit a small part, the feeling became more intense. I took pride in being part of this building, and what it did.

Tuesday, April 27, 1971, the sun rose two minutes earlier than the day before. Dawn broke as I enjoyed my second cup of coffee. I intended to sleep late and enjoy the last day, but I got up early. For such a momentous day, it was rather bleak; chilly for April in Washington, overcast and damp from an overnight shower threatening to last all day. I had looked forward to this day for two years, so, despite the weather, it was a glorious day.

Today was my "wake-up" day. Two years ago, I started with 1095 days, and a wake-up. After the letter from my mother shed light on a fraudulent enlistment, the Army reduced my commitment to 729 days and a wake-up. The last day didn't count as a full day because I would be a civilian before it ended.

Yesterday, I passed everything to Paul, who would face an empty desk across from him for the duration of his service. When I said my goodbyes, I expected as little fanfare as Andy received, but that wasn't the case.

Work stopped at 1600 for the announcement of my departure by the Sergeant Major. The Colonel and the Major added words of appreciation and good luck. I got a hand shake from Mrs. Ramirez, and kisses and hugs from Thelma, Nettie, Pam, and Julie. Louise, the last to hug me, whispered in my ear, "I'll stand in front of you for a minute if you want to hide the boner you got from hugging Pam." Stoner told me on the way home the little ceremony was in appreciation for what I had done, but also recognized I was the first soldier to leave who would not be replaced. It marked the beginning of the end of an era.

"What about Peter and Mullaney? We didn't replace them."

"Peter doesn't count because he didn't end his service here, he transferred. Mullaney didn't do the job you did."

I poured my third cup of coffee after I shaved and put on my uniform for the last time. Les and I had agreed to meet for a last breakfast at the South Post Mess Hall at 0800, after the Pentagon workers cleared out. He'd spent the night saying goodbye to one of the numerous girlfriends he acquired during his time here.

Les and I agreed to report to Fort Myer Personnel no earlier than 0900. Our orders told us to report after 0800, but we decided we would take our time.

The 3.4-mile drive offered no relief from traffic. If my commute at an earlier hour could be described as congested, at this later hour, it could only be called constipated.

As dull as the sky and weather were, the vegetation was bright. Spring came earlier in Virginia than it did in Albany or Boston. The leaves on the trees were full, and the lawns had the brilliant green of fast-growing new grass. Flowers in full bloom announced, "Hey, it's spring. You've been waiting for me all winter. Here I am!"

But, unlike the last eighteen months, I travelled alone in the Sergeant Major's car. He insisted I continue to use it until I left the area. No coffee from the Command Sergeant Major, or my office mate, Ray, who was back to driving for Stoner until August, when he would retire. Ray still faced another year to fight the traffic. I wondered how he'd feel, going from four in the car to driving by himself. No more discussions with the Sergeant Major about history, read about and lived. I enjoyed the solitude but missed the coffee, so I stopped for one at a convenience store at the bottom of the hill on South Walter Reed Drive.

Les was waiting for me at the entrance to the South Post Mess. "Where you been, Gomer? Been here almost ten minutes." His familiar reference to me as Gomer brought the usual smile to my face, as did his exaggerated southern corn-pone accent. "Thunk maybe you re-enlisted so you could enjoy another summer of heat in this arm pit."

"No, Gomer. Just took it easier this morning. Stopped for coffee. Tryin' to slide really smooth into civilian life." We called each other Gomer when we weren't trying to be serious about anything. Today was serious, but a lot less tense than when we first met in the basement on North Post, twenty-one months ago.

We showed our IDs and paid thirty-five cents for breakfast. I had just paid fifty cents for a mediocre cup of coffee at a convenience store. Transition to the civilian world might be costly.

I got an omelet and Les got his grits, biscuits, and gravy, both choosing favorites for our last meal at the best mess hall in the whole of the U.S. Army. As we ate, our conversation became subdued. Les always had a good humor and enjoyed telling me stories of his latest female conquests, or anticipated conquests. At a shade over six feet tall, blond hair, blue eyes, and an engaging personality, he had no trouble getting dates. He still favored airline stewardesses, because they could take him to Florida or the Caribbean for weekends as their guests, where he could maintain his perpetual tan.

In the beginning, while I'd been nervous and apprehensive, he'd been confident. He took the whole Army thing as an adventure. Les was single and unsure of what he wanted to do with his life. He came from money, had graduated college and worked for a big bank in Atlanta, but didn't like it there. The bank job was temporary while he waited to hear on law school applications. When he hit the Washington area, he took it as his playground, and he had his way with it, and everyone in it who wore a skirt. Now, with that confidence gone, he was going back to Atlanta, back to the job that left him longing for adventure.

I, on the other hand, had been dragged kicking and screaming, miserable for having been pulled out of my life when the Army had their way with me.

"Gomer," I said. "You look down today. Thought you'd be up. Getting out. No more boring job. No more uniforms. Good pay. What's up?"

He lost his corn pone, and spoke softly, a serious furrow growing on his forehead.

"Getting out is good. I'm glad for you, because I know you have plans for something better. You say no more uniforms, but that ain't true. I'll still be wearing a uniform, just a different color. A blue suit and a tie. Yeah, my pay will be better, but I didn't have any money problems here. Most of all, I'll be going back to a job I wasn't excited about when I had it. Some folks I worked with back in Atlanta have probably been promoted. Might even have to work for one of those assholes. It's been almost three years since I was in school. Even if a law school took me in September, I don't think I could do the work. It's been too long. Over the past couple of years, I've made a lot of friends here in Washington, and I did more travelling. I'm going to miss my life here."

"Then stay here. Get a job here. Stay with your friends."

"Wouldn't be the same, little buddy. Not as a civilian. I think it's time for me to grow up, and I don't want to grow up."

We finished breakfast and said goodbye to the South Post Mess Hall for the last time. We drove to the North Post, up on the hill, above Arlington National Cemetery, where we would officially sign the papers at HQ Company, U.S. Army making us civilians again.

The day seemed grayer, reflecting Les's mood. When I saw him get out of his car and walk toward me, the confidence in his stride that marked him as special for the last year and a half was gone. He looked like I did when I came into the Army, with my fear of the unknown. His whole demeanor projected lack of confidence; taken with his conversation, I saw it as fear of the known.

We parked in the lot on the side of the old red-brick building that served as the Personnel Center for Fort Myer. The building housed the services and personnel who worked behind the scenes to make sure all the soldiers stationed at Fort Myer and the Pentagon got housed, fed, and paid. Until sixty days ago, Les, Dexter W., and I were unknown to anyone other than the Finance people at Fort Benjamin Harrison. The IBM cards I kept hidden under my blotter for the last eighteen months made us invisible.

Dexter W. had entered the Army the week before we did, and went through this check-out process the previous week. Dexter W. told us mustering out was very anticlimactic. Nothing like that first day at Basic Training. Just sign some papers, get some papers, and done. No one said, "Thanks for coming," "Hope you had a nice time," "Keep in touch"—nothing, just a reminder to register with Selective Service as veterans.

Most of the reason the checking-out process went quickly and smoothly was because the Army gave us a month to do a lot of pre-checking-out things. We had our final physical, both medical and dental, completed during this time, as well as some routine paperwork, so it was in place, with an effective date set for our last day, our wake-up.

Before our last day, we had to turn in clothing. In the real Army, the soldiers spent most of their time in their fatigues. The Army figured these would be shot and didn't ask us to return them. Same with the combat boots, underwear, and socks they issued us. The soldiers who worked the Pentagon wore their Class A uniforms or their khakis all the time. But, in the one-size-fits-all mentality of the Army, we could keep our fatigues, but needed to turn in our Class A uniforms and one set of khakis. Some guys argued, and won, that if they turned in their uniforms, they'd have to wear dirty or wrinkled clothes to work. The guys at Fort Myer didn't care, and let us win our weak arguments, except for two items.

Two pieces of clothing could transition into civilian life easily, a raincoat and a field jacket. The raincoat was a belted, off-gray that would work in the civilian work force, or the grunge of a college campus. The field jacket was the uniform of the war protesters, and the preferred jacket for hunters, campers, and backyard mechanics. It was also the one piece of clothing the guys at Fort Myer wouldn't overlook; it must be turned in. They heard all the excuses and accepted none. If you said you didn't have one, they told you to buy one at the PX to turn in. I reluctantly turned in my field jacket.

The Army didn't make us turn in our Army "great coat," designated "Overcoat AG44" on our clothing list. I never wore my coat, but I remembered pictures of World War II G.I.s wearing the "great coat" in Europe. The long coat, with the collar turned up, covered the body from the ears to the ankles. It was heavy because of the dense, waterproof outer material, and made even heavier with a liner. Not a run-of-the-mill liner, but as heavy as the coat, extending into the sleeves and covering the inside of the coat to the ankles. They let us keep these coats because we were still within the Army's definition of cold weather, Washington. Again, using the Army "one definition fits all," the cold-weather season for Washington was the same as Nome, Alaska—that is, September 1 through May 2. I guess they didn't want us to catch cold going home, so they let us keep the "great coat."

Dexter W. overestimated the drama of mustering out. It was less than anticlimactic. No one yelled at us as they did when we came in. There was no urgency in moving us from one desk to another. To the people attending to us, this was routine. All the soldiers who came into the Army two years ago were now getting out; nothing special.

This bothered me, because I was used to being special and being treated special. I worked at the Pentagon and had "Third Herd" shoes! DeWitt Hospital and Rader Clinic treated me special! I liked being special. As a civilian, I'd have to adjust to being just another "used to be somebody."

We took about an hour to sign out: no rush, no waiting, just good practice at not being special. We moved along to Finance and picked up our last check, again signing papers and getting more receipts. They provided us with an address of a Selective Service office in Arlington, Virginia, to register as veterans.

And then it was over.

Les and I prepared for the lack of drama and ceremony and had decided we would have our own ceremony. When we entered the Army on that first day at the AFEES station, we had to take a step forward to become soldiers.

Les said, "Prepare to become a civilian." As the people around us watched, we stood at attention. Les said, "Take one step back to be a civilian." We did, and we were.

In a final gesture to Peter, Andy, and the guys who preceded me in leaving 1D726, I said, loud enough to be heard by all, "Let's bust out of this puke hole," and we did.

More rain was on its way as I pulled into the Selective Service parking lot. Les got out of his car when he saw me drive in. He had withdrawn even more. His broad shoulders, once square and thrust back proudly, now had a forward slope, making them appear rounded. His head, always held high, thrusting his chin forward, now leaned forward, forcing the appearance of a double chin. The blue eyes, once bright and full of mischief, were now dull.

We joined the crowd who had gotten there early. We were out of the Army, but now we had to hurry up and wait in a civilian line. If patience is a virtue, why is waiting such a pain in the ass? At least we didn't hurry-up as early as the guys in front of us, and we wouldn't have to wait as long, because we gave them an hour head start.

We registered as veterans, and were done.

I smiled, Les was dour. I suggested a beer to celebrate. He declined, saying he was going to try to make it to Atlanta that evening. Rain started to fall, but gently. The rain smelled of spring, a renewing rain, and fit my mood. We shook hands.

"Let's stay in touch," said Les.

"Will do, Gomer," I promised.

Les and I had not shared phone numbers or addresses, but made the same promises thousands of other soldiers had made for years. We made the same promise men and women had made for hundreds of years after a one-night stand.

As I watched him walk to his car, I saw a man in conflict with his future. I saw a man whom I admired for almost two years. A man I now felt sorry for.

I never saw or heard from Les again.

When I got to my car, I saw a happy face reflected in the glass of the car door. A face that had grown in confidence and experience over the past two years. A face looking forward to the future.

I got in the car, lit up a smoke, and thought about the journey I had taken—a journey that started with a bus ride.

On Friday, Nixon announced the last Marine combat units left Vietnam. With me and Andy out of the Pentagon and the Marines out of Vietnam, I didn't like our chances anymore.

***

Postscript note: Six weeks after I got out, the New York Times began publishing the Pentagon Papers. All the secrets I learned or suspected during my time in the Five-Sided Paper Factory were now public knowledge. Returning to school, a veteran who served at the Pentagon, I was suspect as a government undercover agent. The myth grew when I was the only one not shocked when the Watergate scandal broke. I was less shocked at Watergate because of the experience I'd gained learning Mr. Ellsberg's secrets on my own. Why wouldn't the Republicans be spying on the Democrats if the Army was spying on itself? My response to my fellow medical students, outraged by the "plumbers," was simple. "The call to the police was the second call made by the DNC. The first call was to their version of the plumbers, telling them to remove the bugs they planted in the RNC Headquarters."

Postscript 2: On January 6, 1972, when winter had a firm grip on New Haven, with temperatures hovering around fifteen degrees, I was glad the Army kept the field jacket and forced me to take the great coat. The coat covered me from ears to ankles. Those hippie wanna-bes who were practicing to be protesters wore the Army-Navy Store field jackets, and froze. Those who asked me where I got my neat coat were puzzled with my answer: "The Five-Sided Paper Factory."

AUTHOR'S NOTE

Dear Reader,

Thank you for reading The Pentagon Years. I hope you enjoyed it. If you could take a minute to write a review for it on Amazon or Goodreads, or better yet, both, I would be very grateful. Reviews, as well as word of mouth referrals are very important to writers. Authors live for reviews, sales are good, but reviews help generate sales when people like you share your enthusiasm for a book. Being an independent one-man show, any help I get spreading the word helps and I thank you in advance.

Writing a review is easy. On Amazon type in William J Kennedy The Pentagon Years. Click on the book cover and scroll to the end of the section where you'll be invited to write a review.

If you enjoyed The Pentagon Years and would like to know what happens to Jonathan West when he leaves the Pentagon, I invite you to read First Kill and Morally Gray, more serious thrillrers. The fourth book in the series is in concept development now and should be available in 2020. Follow me on Facebook at William J Kennedy Author, or drop me a line at consultkennedy@aol.com.

I've enjoyed writing for you. I hope you've enjoyed reading.

ABOUT THE AUTHOR

William J Kennedy is a retired pharmaceutical executive currently living in Lewes, Delaware. He can be contacted by email at consultkennedy@aol.com.

www.ingramcontent.com/pod-product-compliance
Lightning Source LLC
LaVergne TN
LVHW091212080426
835509LV00009B/956